HOW SOVIET FOREIGN POLICY FAILED

What Complexity Science Tells Us That Nothing Else Can

HOW SOVIET FOREIGN POLICY FAILED

What Complexity Science Tells Us That Nothing Else Can

Robert M. Cutler

EMERGENT™
P U B L I C A T I O N S

The cover image "Ball Fracture 1993" was created by the author under Windows 3.1, using the pioneering shareware image manipulator Picture Man, described as being "designed for true color image enhancement, retouching, painting, and the creation of special effects" but which its creator Dr. Igor Plotnikov appears to have orphaned as an 8-bit app without progeny.

How Soviet Foreign Policy Failed
What Complexity Science Tells Us That Nothing Else Can
Written by: Robert M. Cutler

Library of Congress Control Number:
    2013946130

ISBN: 978-1-938158-11-7

Printed in the United States of America

Καλουμενη

CONTENTS

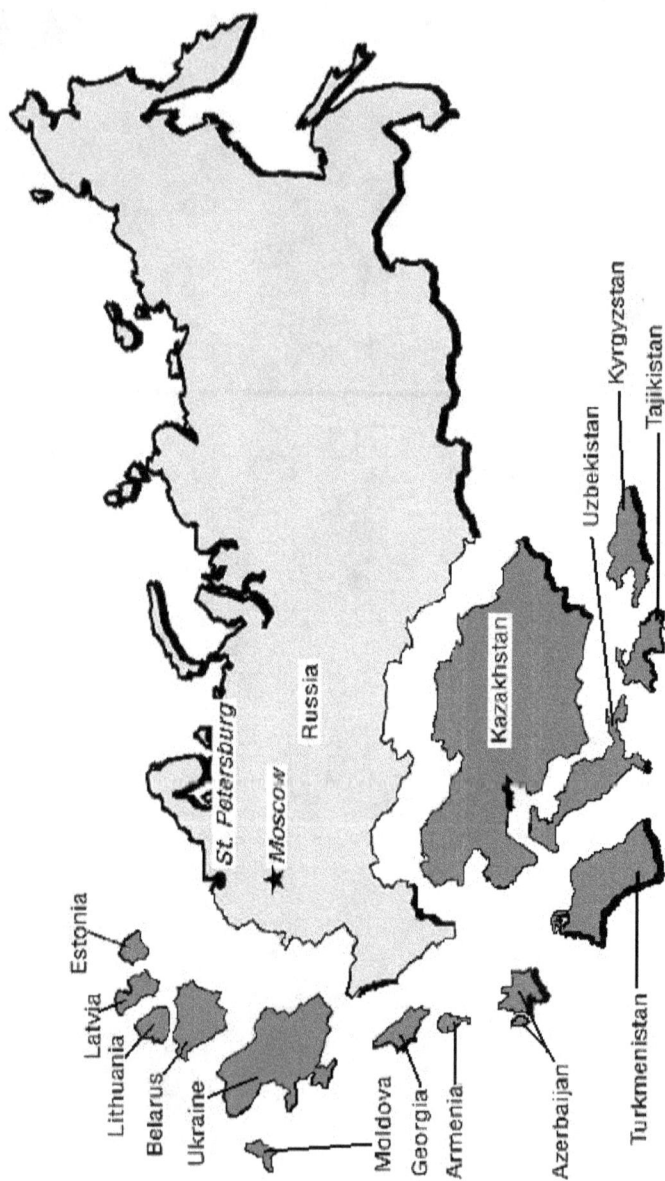

Introduction

[The philosopher of science Ernst Cassirer] discloses the basic character of science as the eternal attempt to go beyond what is regarded [to be] scientifically accessible at any specific time. To proceed beyond the limitations of a given level of knowledge the researcher, as a rule, has to break down methodological taboos which condemn as "unscientific" or "illogical" the very methods or concepts which later on prove to be basic for the next major progress.

— Kurt Lewin (1949: 275)

The disintegration of the USSR could be considered one of the greatest management failures of the twentieth century. However, it has rarely been treated from this standpoint. A notable exception (Naylor, 1988) concluded, at a time when many Western observers were still skeptical, that Gorbachev's reforms were real because they were necessary for the Soviet Union to survive. This chapter examines a little-studied aspect of the Soviet political system, as an index to ascertaining just why those reforms were unsuccessful. That aspect is the overlap between making foreign policy and making foreign propaganda. Indeed, although the Soviet press received special attention among Western analysts of Soviet behavior, the general topic is not one that is restricted to communist countries (e.g., Cohen, 1970; Wittkämpfer & Bellers, 1986). For the Soviet press was part of the Kremlin's central nervous system that not only moved in the external world but also absorbed and interpreted information from it. If the Soviet Union did not survive, how it processed information about its environment is an important key to understanding why.

Western specialists on the Soviet Union during the Cold War encountered, without realizing it, many issues under current discussion in management science. Those issues presented themselves as problems in understanding the cognitive aspects and organizational development of the Soviet political system. The political disintegration of the USSR can be regarded as a failure by the Soviet system to adapt successfully to demands from increasingly complex international and domestic environments. As such, there is direct relevance to the situation encountered by managers in complex bureaucracies today. Indeed, the first major political interpretation of the post-Stalin Soviet system (Meyer, 1965) literally characterized it as "USSR, Inc.," in order to make the point that it was organized as a large bureaucratic institution. The reasons why the USSR failed to adapt, and also the reasons why Western analysts failed to foresee the entire situation, are therefore of interest. This monograph discusses the organizational and cognitive evolution of the Soviet foreign policy making Establishment in particular. It also traces attempts and failures by Western Sovietologists to understand those changes in the structure and cognition of the Soviet foreign policy making Establishment. It further addresses the question of Soviet organizational learning in that context, and draws lessons for contemporary complex organizations, including business corporations.

The traditional point of departure for the study of Soviet domestic and foreign policy was the totalitarian theory. Useful as this was during the Stalin era, however, it did not begin to capture the evolving complexity of the post-Stalin system. According to that theory, no organizationally-based explanation of the mechanisms of Soviet foreign policy making was necessary, because a single set of immutable rules—an "operational code" (George, 1969; Leites, 1951)—was said to prescribe Soviet foreign policy behavior. That behavior, in turn, was assumed to be highly deterministic, wholly unreactive to external stimuli, fully resistant to change, and therefore also incapable of learning (Friedrich & Brzezinski, 1956). The study of the Stalin era supplied the original parameters for specifying the totalitarian model to Soviet conditions. Implicit in this theory, for which Stalin's regime was the evidence, was the assumption that the system did not allow any competing interests.

However, the succession to Stalin did not conform to the totalitarian theory, according to which a new dictator should promptly consolidate power and maintain it unchallenged. The totalitarian theory's explanation of political conflict—that it occurred during succession crises but only then—became implausible after 1955, when Khrushchev and Bulganin established themselves as a duumvirate. The study of the succession to Stalin, and the resistance of the Stalinists to Khrushchev even after Khrushchev's 1956 de-Stalinization speech, was what led Western Sovietologists to elaborate a new theory called the "conflict-school."

The totalitarian theory held that the will of the totalitarian dictator determined foreign policy and that therefore no organization played any important role in Soviet foreign policy making. By contrast, the conflict-school theory recognized the significance of one organization: the Politburo. However, it focused upon the Politburo not as an organization *per se* but only as a forum for conflict among its members. The conflict school considered the struggle for political supremacy to be the only domestic policy process relevant to foreign policy making, but it considered that this struggle was a permanent characteristic of Soviet politics. According to the conflict-school theory, the one significant organization (the Politburo) did not exhibit any behavioral regularities. This was an advance over the totalitarian theory from the standpoint of organization theory, but still the conflict-school research program imputed to the Politburo a total lack of routinization.

The increasing complexity of Soviet political communication after Stalin's death

Since access to means of communication was always a scarce political resource in the USSR, the post-Stalin Soviet media always attracted people seeking to influence policy. Informal links between press organizations and opinion groupings arose in an *ad hoc* manner. When members of different groupings acquired executive responsibility in different newspapers, they and their associates aggregated into loose cliques around the editorial boards. It became known in an unofficial way that different climates of opinion predominated in different journalistic circles. Disagreement or conflict among them—still informal— occurred and found public expression. An editorial decision to publish an article in a Soviet newspaper did not need Central Committee approval. This could be done on the authority of a newspaper's editor-in-chief or international affairs page editor, who could even request nonstaff contributors to the newspaper to prepare such articles. This was frequently part of a conscious attempt to generate a heterodox policy alternative or defend an orthodox one: an act of group advocacy.

The Soviet press under Lenin was characterized by liveliness and experimentation, but new controls were placed upon the press and censorship was reinforced with the rise of Stalin. The Stalinization of the Soviet press continued through the upheavals of the 1930s. The late 1930s in particular witnessed a systemic transformation of the organization of the Soviet press very soon after the most spectacular purge trials concluded (Fainsod, 1956; Finn, 1954; Fogelevich, 1937; Gaev, 1953a, 1953b, 1955; Kotlyar, 1955; Shulman, 1948). Wide-ranging transformations of the Soviet press system as a whole took also place in the mid-1950s and again in the early 1970s, soon after the establishment in power, respectively, of Khrushchev and Brezhnev as the political leader first among equals (compare Révész, 1974).

Khrushchev, upon defeating the "Anti-party Group" of Stalinists within the Politburo (Pethybridge, 1962), reestablished the Union of Journalists so as to modify the top-down function of the Soviet press and increase the relative autonomy of people working in the press (Remington, 1985a). He greatly increased the number of newspapers and journals in order to reach differentiated strata of Soviet society (Roth, 1982: 175-206). This development gave those readerships an opportunity to communicate with more specialized editorial boards, providing a higher quality of feedback on specific issues from the society at large to the political elite. As the number of literate journalists increased after 1956, along with the number of institutions and newspapers in which they worked, the entire system began to become more complex. As the sophistication of working journalists increased, editorial boards were also given greater prerogatives to decide what could be published. (To give one minor but indicative example: Under Stalin, the next day's copy of *Pravda* was reviewed inside his personal secretariat before publication [Rosenfeldt, 1978]. Under Khrushchev it was no longer possible for central directives, even working through the Propaganda Department of the Central Committee, to control in advance everything that was published in the Soviet press.)

The changes in the late 1950s had an overall reformist effect on the channels of political communication. Those of the early 1970s had an overall conservative effect. In particular, the intent during the first period was to increase the number of roles in press organizations as well as their associated prerogatives, and to expand the openness (especially bottom-up vs. top-down) of the networks political communication in which they functioned. The intent of the latter period was not so much to reduce these as to rationalize them. "Organizational complexity" could therefore be said to have begun to increase in the Soviet press and the Soviet system at large, starting in 1956: so much so, that after Khrushchev was deposed in 1964, the new Soviet leadership convened another Congress of Journalists to deal with the aftermath. But the tide could not be turned back to the rigidly institutionalized lines of political communication of the Stalin years.

As it was put at a closed meeting on problems of managing press organizations, by the Party's official representative on the editorial board of *Pravda* responsible for giving the newspaper its political direction (called its "responsible secretary"), editorial organizations now depended upon "manifold sources" of information. These included social organizations, scientific establishments, cadres in the central apparatus, and specialists from outside the newspaper staff, in addition to the traditional sources of political direction from the Central Committee and its own staff correspondents (Tsukasov, 1973: 15). A

Soviet expert on the organization of Soviet press—a Moscow University journalism professor—confirmed the need for a new conceptualization, and in passing also confirmed the intersection of the policy-making and propaganda-making systems (Shkondin, 1982: 61, highlighting in the original):

> The aggregate of editorial organizations *is also a component of the system of the press. Each* [editorial organization] *develops around a press organ and includes the creators of literary works and texts that are subject to circulation and distribution among a mass audience. The editorial organization ... is a significant factor of the formation and realization of policy, and actively influences the working out of decisions in the system of the political institutions of society.*

His analysis of the relationship between the Soviet press system and the Soviet policy-making system even included an intimation of an attempt to move towards complexity theory. However, he still remained constrained within a framework of general systems theory (Shkondin, 1982: 62-63, highlighting in the original):

> *Any social system, including the press, requires organization and management. These affect not only* the organs of editorial management *but also* the organs of the party[-oriented and party-animated] political direction of the press... *The basic components* [of the press] *are its* [sub]*system of direction* ['rukovodstva'] *and (sub)system of editorial management* ['upravleniia'], *editorial organizations,*

publications, and audiences. Since their functioning is subordinate to the realization of the press as a system, each of the [just-]*named* [subsystemic] *components includes not only structural formations but also processes that occur within and between* [each of] *the given subsystems, as well as between* [each of those subsystemic] *components and the environment.*

Empirical work by Soviet sociologists of the Soviet press, generally ignored by Western Sovietologists, further reinforce the deduction (Cutler, 1984b, 1985a, 1985b) that the editorial staff and of the nonstaff opinion-circle (in Russian, *kruzhok* or "little circle") around the editorial board became salient determinants of the opinions published in Soviet newspapers (Sekerin, 1973: 61):

The "publisher" who defines the program of activity of a newspaper [i.e., in the Soviet system, the bureaucratic institution that publishes the newspaper and is politically responsible for it] ... *as a rule chooses the main task or group of basic tasks. In recommending the discussion of whatever question or in placing whatever task before the editorial staff, the "publisher" frequently ... disaggregates the problem, and these most significant ideas (instructions, goals, etc.) will in one way or another be repeated in the concrete publications.*

...[H]*owever, the editorial staff realizes the task provided by the "publisher" not in a "pure" form but in the form that most suits its own point of view. This* [intervention by the editorial staff] *may make the idea-task more intelligible to the reader but, being an aspect introduced internally* [from within the newspaper itself], *it may also sometimes amplify aspects of the idea that impede or even deform the reader's perception* [of the "idea-task" intended by the newspaper's "publisher"].

Such a "deformation" may be manifested through a "form that most suits [the editorial staff's] point of view." This means that the editorial staff may introduce policy views that do not uniformly accord with those of the "publisher"; and as a deputy editor-in-chief of Tass (Fadeichev, 1971: 189) has explained, "The editorial staffs of newspapers ... sometimes publish different versions of the account of one and the same event, ... and draw their own conclusions and commentaries, giving their own evaluation of an event [, and they] have the full right to do this."

From hierarchy to complexity in the Soviet foreign policy making establishment

Once the conflict-school theory admitted the permanence of intra-elite conflicts as a matter of principle, it became inevitable that some analysts would begin to look for *institutional* bases of regular patterns of elite conflicts. Scholars in Soviet area studies also began to realize that post-totalitarian modernization led to "feedback" from Soviet society to the political system (Löwenthal, 1970). Reforms in the Soviet domestic propaganda institutions confirm this (Hoffmann, 1968). Analysis thereby shifted from the struggle over personal power to that over policy substance. Rivalries among institutions became the fodder for analysis. Policy, not persons, became more and more the focus of study. The aggregation and articulation of demands moved closer to the center of their attention, and also became easier to study, because the raw materials for analysis-such basic items as Soviet newspapers and other publications-became more available in increasing amount. As analysts tried to understand these new political demands and especially who made them, they concentrated on institutional groups, particularly those defined categorically by occupation (Lodge, 1969; Skilling, 1966; Skilling & Griffiths, 1971).

The institutional-group theory, typified by Aspaturian (1966, 1972), evolved from the conflict-school theory. Aspaturian distinguished between Soviet interest groups having domestic organizational goals leading them to benefit from international tensions, and those having such goals leading them to benefit from the relaxation of international tensions. These groups he defined occupationally. For example, the military was assumed to prefer tension because this increased its budget, whereas light industry was assumed to prefer the absence of tension because it wanted part of that budget. This theory analyzed Soviet policy making to see how those institutions formed coalitions domestically to pursue foreign policy goals that they had in common still satisfying domestic constituencies. According to the institutional-group theory, conflict occurred in Soviet foreign policy making not only among leaders and the institutions they ran but also among those institutions independently of leadership conflict.

The institutional-group theory also derived from the so-called "interest-group" approach to Soviet politics. This approach was introduced into the study of Soviet politics in response to the need for an analytical construct that could be applied to systemic conflict over specific issues of immediate policy relevance. Because such issues touch directly on organizational goals, the premise that institutional bureaucracies have unitary interests became an implicit assumption of the institutional-group approach. The institutional-group theory of Soviet foreign policy making arose as Sovietologists adopted the analytical distinction between conflict resolution within the elite, and interest aggregation and articulation throughout the broader political and social system. This distinction produced the bifurcation in Sovietology between the study of group and elite attributes and that of group and elite activity.

Analysts of group and elite attributes (Blackwell, 1972, 1973; Fleron, 1969a, 1970) attributed great significance to the occupational-category variable, partly because of a tendency to look at aggregate statistics on career paths. This focus led interest groups to be explicitly equated with the institutions into which the Soviet system organized categorical occupational groups. That conception is what generated the institutional-group theory of Soviet foreign policy making. Indeed, this connotation of *interest group* was useful because it satisfied the need for an analytical construct that would help explain how opposing policy goals could be produced from within the Soviet system. In fact the interest-group approach also generated three separate techniques. It is even more useful to distinguish unambiguously among all four:

1. The *interest-group approach* (Skilling, 1966) was originally a primitive technique for grasping the political sociology of the Party, and, as such, it tended to deal with specific policy issues rather than with broad predispositions.

2. The *specialist approach* (Solomon, 1978) stressed the participation in the process of agenda-setting, policy-making, and policy-execution, by individuals having specialized "technical" knowledge.

3. The institutional-pluralism approach (Hough, 1972) related to the Soviet political process generally, and to broad policy predispositions rather than

to particular decisions; indeed, when the general interest-group technique was applied to the analysis of foreign policy making, it became limited by its own implication, which institutional-pluralism approach made explicit, that institutions were unitary and monolithic, with unambiguous interests (Skilling, 1983: 10).

4. Finally, *tendency analysis*, which was developed with specific reference to foreign policy analysis (Griffiths, 1971, 1972) concerned broad philosophical principles that individuals in different institutions have in common, focusing as much on the content of the shared images as on the individuals sharing them.

The tendency-analysis technique generated a field/ground shift that focused not on the institutions as harboring individuals having ideas but on the ideas themselves, and which treated the individuals merely as their carriers. From this there developed the "interactionist" theory of Soviet foreign policy making, which focused not on the institutions but on the shared cognitions within organizations. The interactionist theory provided tools for discovering that organizations could be fragmented, and that different parts of the same bureaucratic institution could have different images of the international environment. From this it followed that if the institutions themselves were not assumed to be unitary, then coalitions could bring together parts of different institutions. The interactionist theory recognized that these institutions received more information, but it did not imply that they required more differentiated processing according to a cognitive hierarchy. It implied that the USSR could learn international behavior at both organizational and individual levels, and that Soviet foreign policy was very voluntaristic and responsive to the general international climate.

Thus the interactionist theory implicitly regarded Soviet foreign policy behavior as the output of a system that processed information on the international situation. It emphasized what happens to information between the cognition of an external situation and the choice of a policy in response to it (Axelrod, 1972, 1973). Tendency analysis was principally concerned with cognitive content; the interactionist theory added an organizational context. It drew attention to how organizations formulating foreign policy transform information, and to how they influence the ways that individuals interpret information. The interactionist theory implicitly assumed an increase in the volume and variety of available information compelled every organization to construct a belief system. It also implicitly assumed that the organizational complexity of the post-totalitarian Soviet system permitted each organization to draw on its own resources to that end. It recognized that in the Soviet system there were more people, there was more knowledge, and there were more expertises. The result of this proliferation was that people inside institutions became able to find allies inside other institutions instead of being restricted to staying politically inside their own institutions. The result was that the articulated interests were no longer hierarchically organized by the systemic structure of the institutions; rather, the interests began to self-organize on sub-institutional levels.

The interactionist theory was a fundamental advance over the institutional-group theory, according to which the highest political leaders reflected unitary institutional interests and their competition imposed operational codes on those institutions that limited real differences to a question of preferences for or against international tension. However, like the institutional-group theory, the interactionist theory conflated general predispositions with attitudes on specific foreign-policy issues. The interactionist theory did explicitly provide (unlike the institutional-group theory) for the existence of competing "operational codes" in the Soviet foreign policy making Establishment, and even affirmed implicitly that competing operational codes could exist inside individual institutional bureaucracies. However, it did not—and could not—affirm that complexity separated general predispositions from particular policy preferences as distinct and interactive levels of cognition and information processing within and between organizations. For that theoretical advance to be made, it was necessary not only to treat foreign policy making explicitly as information processing about the international environment, but also to integrate the various different approaches to the study of interest articulation (the interest-group, tendency analysis, institutional pluralism, and specialist approaches) that had developed.

Because the concepts of complex systems did not exist at that time, this step was never taken. As a result, the basic ideas that motivated the interactionist theory were never systematically integrated in a comprehensible form. The crucial step involved distinguishing systematically among the various techniques for studies interest aggregation and articulation in Soviet politics that were enumerated above. Each of the four approaches carries assumptions about the type of interest-bearing aggregate that is assumed to exist and about the scope of policy that is analytically relevant. The aggregate to which an "interest" is imputed may be either an institutional bureaucracy or a functional collection of groups and individuals-a "grouping"-that cuts across institutions. The scope of policy may be either a general predisposition toward political life in general, on the level of instrumental operational-code beliefs, or a particular issue on which there are well defined and contrasting positions in specific terms, on the level of cognitive-map elements. Table 1 takes this integrative conceptual step.

Thus if the interactionist theory largely failed to gain adherents in the 1970s and 1980s, this was also because political science had not yet digested enough cognitive and social psychology to provide the vocabulary necessary for expressing systematically the conceptual insights (compare Breslauer and Tetlock, 1991). Moreover, cognitive and social psychology had not elaborated the full conceptual framework to provide the appropriate analytical techniques, including rigorous methods of discovery and analysis. In much the same way, only the present-day context of complexity theory allows us to understand more fully how the Soviet foreign policy making Establishment evolved from the 1950s through the 1970s and beyond into the Gorbachev

		Type of Attitude Selected as Analytical Focus	
		General inclinations and cognitive predispositions	Particular opinions about specific decisions
How the Actors are Defined	As bureaucratic institutions	*Institutional pluralism* (e.g., Hough, 1972)	*Interest-group approach* (e.g., Skilling, 1966)
	As functional aggregates transcending and splitting formal organizations	*Tendency analysis* (e.g., Griffiths, 1972)	*Specialist approach* (e.g., Solomon, 1978)

Table 1 *Classification of approaches to the study of soviet policy-making.*

era. Only the systematic integration of the interactionist theory's implicit advances in organizational analysis with the principle of the cognitive hierarchy could produce a theory sophisticated enough to grasp the intricacies of the Soviet foreign policy making Establishment as a complex organization (Cutler, 1990). Let us therefore call such a theory the "complex-system" theory of Soviet foreign policy making. The next section of this chapter elaborates this theory and distinguishes it from its predecessors.

Complex-system theory in a Soviet context: A structural-normative clarification

Subsequent theoretical developments will be more comprehensible if we briefly to put these developments just described into an epistemological perspective. This perspective is drawn from Winch (1958). According to him, a proper theory of knowledge for social-scientific research includes the definition of structures, norms, and behaviors.. The original decision-making framework in political science (Snyder, Bruck & Sapin, 1962) drew attention to three "clusters" of variables or sets of phenomena. Each addressed some facet of how people making decisions in organizations operate. These were "spheres of competence," "communication and information," and "motivation." In an organizational-science context these are respectively equivalent to authority in organizations, communication in organizations, and goals in organizations. The first decision-making cluster, authority in organizations, is about organizational *structures*.

The second decision-making cluster, goals in organizations, is about organizational *norms*. Here that means the presence or absence of a "cognitive hierarchy." The presence of a cognitive hierarchy means a model of cognition and choice that defines doctrine, ideology, strategy, and tactics as interrelated but distinct ranks of information processing (Abelson, 1973; Heradstveit & Narvesen, 1978; Peffley & Hurwitz, 1985).

The absence of a cognitive hierarchy means the levelling of information processing to a single cognitive rank, where no concept is more general or more specific than another (e.g., Converse, 1964). (Winch points out that among structure, norms, and behavior, any two may be properly used in tandem to study the third. The third decision-making cluster, communication in and among organizations, is about *behavior*. An organization's behavior in this instance is its articulated interest on a foreign policy issue.)

More about norms

A belief system consists of an integrated hierarchy of cognitions differentiated by the level of their abstraction. It is convenient to specify four levels in such a belief-system: moving from the abstract to the concrete they are "doctrine," "ideology," "strategy," and "tactics." Doctrine refers to immutable historical destiny defining the basic actors, their relations, and how these will turn out. Ideology concerns the particular combination and succession of themes of conflict and cooperation (Cutler, 1982c). Strategy concerns overarching patterns of conduct animated by those themes. Tactics are situational moves designed to achieve specific and immediate goals, the succession of which forms a strategic pattern. In specific examples of Marxist-Leninist reference:

1. ***Doctrine** is the immutable teleology* dictating how to define the dramatis personae of international politics and plotting the script of their theater. For example, the conflict between the Soviets and

the Chinese in the late 1950s and early 1960s was a dispute over *doctrine*, because it concerned the number and the composition of different "camps" in world politics, and whether the use of nuclear weapons could hasten the victory of world socialism (Zagoria, 1962).

2. ***Ideology*** *consists in the ensemble of idea-elements* defining the categories of cognition. It is the vocabulary that expresses eternal doctrinal truths. The Soviets called this ideology "creative" Marxism-Leninism, because its idea-elements could combine in different ways.

3. ***Strategy*** *is the goal-oriented line* of action prescribed by ideology. When the categories of ideology concatenate to define what really exists, they limit the set of possible futures because what is inexpressible in their language becomes *a priori* impossible. Specific conceivable futures imply goal-oriented lines of action to attain them, and each goal-oriented line of action is a strategy.

4. ***Tactics*** *operationalize strategy* at the level of actions, intentions to act, and preferences for actions to be taken, with reference to concrete issues in particular policy fields.

George F. Kennan (1967: 485) remarked in the late 1940s how struck he was by "the importance, in the formulation of national policy, of two things: first, one's idea of one's own country, its capabilities, and its natural role in the world; the other, the interpretation given to the psychology, the political personality, the intentions, and

27

the likely behavior of the adversary." Empirical belief-system analysis calls these two dimensions "image of Self" and "image of Other." These are dimensions of a formalized operational code. In slightly more social-scientific language, therefore, we may say (see Table 2): doctrine is absolute revealed truth; ideology corresponds to philosophic operational-code beliefs; strategy corresponds to instrumental operational-code beliefs (George, 1969); and tactics correspond to elements in cognitive maps (Axelrod, 1976). From the interactionist and complex-system theories it followed that different organizations acquired *different* operational codes, influencing their preferences on specific foreign-policy issues. Traditionally, Western scholars simply referred to the "Left" and the "Right" in Soviet foreign policy. These were appropriate labels, because they descended directly in a line, respectively, from the Left Opposition and the Right Opposition to Stalin (Daniels, 1960; Dallin, 1981). In information-processing terms, the Left operated on the basis of an ideologically orthodox model of the international environment, while the Right operated on the basis of a more heterodox one.

Altered perceptions caused changes in Soviet attitudes towards world politics in general and towards other international actors in particular. The following example from the Brezhnev period illustrates exactly how this occurs. It prepares the way for a deeper discussion of how the complex-system theory of Soviet foreign policy making treats learning. For a long time after World War II, Marxist-Leninist ideology gave pride of place in the capitalist world to the United States, and treated

Level of Cognitive Hierarchy	Empirical Referents (Axelrod, 1976; George, 1969; Leites, 1951)	"Conceptual Dependency" Level (Abelson, 1973; Cutler, 1982c)
Doctrine	Immutable Historical Truth (e.g., dramatis personae of history)	Script
Ideology	"Philosophical" Operational-Code Beliefs (e.g., the essence of life as conflict or cooperation)	Theme
Strategy	"Instrumental" Operational-Code Beliefs (e.g., maximum vs. minimum plans)	Plan
Tactics	Cognitive Maps	Molecule

Table 2 *Levels of the cognitive hierarchy and their associated names in various different schemata.*

Western Europe and Japan as appendages to "American imperialism." However, by 1970 a large number of Soviet experts on international affairs regarded the European Economic Community (EEC) as an economic "center of imperialism" equal in stature to the United States. A new "theory of the three centers" of imperialism proposed that, at least economically, Western Europe and Japan had freed themselves from American tutelage and actually competed with Washington for influence within the international capitalist system (Mel'nikov, 1972; cf. Adomeit, 1979; Hough, 1980). Soviet attitudes towards West European integration were also strongly affected by the EEC's response to the upheaval in the international monetary system in the early 1970s (Kozlov, 1975). The authoritative Soviet evaluation (Anikin, 1971, 1973) of the practical significance of the EEC's economic integration depended upon how serious its political coordination was. Monetary integration in the EEC was taken as an indicator of the seriousness of that political coordination, and the key to the Soviet controversy over EEC monetary integration was the seemingly esoteric issue of the role of gold in the world capitalist economy (Rémy, 1981).

The orthodox tendency in the Soviet foreign policy establishment (e.g., Stankiewicz, 1972) emphasized the economic, political, and especially military threats arising from the EEC's consolidation and expansion. An extreme orthodox view (e.g., Gantman, 1972) denounced West Germany's desire for a supranational institution to regulate the proposed EEC currency system as an expression of renascent German imperialism. Proponents of this view (e.g., Bezymensky, 1972) could conceive a unified Europe only as a NATO puppet manipulated by a remilitarized West Germany. Some of them (e.g., Stadnichenko, 1971) even thought that West Germany sympathized with supposed American attempts to torpedo the formation of an effective EEC monetary union. The heterodox tendency in Soviet foreign policy did not interpret EEC monetary integration in this way. According to it, the FRG sought supranational institutions simply because "it fears that without [them] it will be obliged to pay for the inflation in other countries of the Community" (Zhdanova, 1971: 86). Heterodox analysts (e.g., Bolotin & Kudrov, 1972; Gromeka, 1974) recognized that Western Europe's economic base permitted it to emerge as a "center of imperialism" independent of American tutelage. They saw the creation of an EEC monetary policy as a political manifestation of the EEC's irrevocable place in world affairs. Further, they argued (e.g., Stepanov, 1972) that such a degree of political integration would not lead inevitably to military integration.

The "theory of the three centers" was nothing less than a revision of the *ideology*. As such, it had important implications for the *strategy* of the international communist movement. If world capitalism had three centers, then socialist revolution might occur in one of them (such as Western Europe) long before it did in another (the United States or Japan). Various strategies could promote to this. For example, the slogans "united front" and "popular front" are drawn from the history of the Communist International between the two world wars. They summarize the possibilities for communist cooperation with other political parties (Kukhtevich,

1972). The first is more orthodox and consummatory, the second more heterodox and instrumental. The united front entails an exclusionary "three-class alliance" strategy comprising only workers, peasants, and leftwing radicals. The popular front is an inclusionary "four-class alliance" that admits also certain strata of the bourgeoisie into the coalition. Historical experience links the united front and its three-class alliance to the image of revolution as cataclysm, a maximalist image in which revolution is foremost a political act and great strides are made through a limited number of earth-shaking changes. The popular front strategy emphasizes numerous minor tremors in the political landscape, reflected in incremental social change.

Thus the united-front and popular-front policy tendencies animated an internal Soviet policy dispute in the early 1970s over the proper strategy for West European communist parties. The codewords "socialism" and "democracy" respectively expressed the united-front and popular-front tendencies. To pursue "socialism" meant accelerating the near-at-hand collapse of capitalism and catalyzing the political revolution that would bring its final downfall. To pursue "democracy" meant emphasizing social change and promoting it gradually, through such measures as extended economic planning, nationalized industry, and enhanced workers' control over economic production. One of the clearest Soviet analyses (Zaretskii, 1973: 109, 110-11, 119) put it this way:

> *Taken to its logical end, the struggle for democracy throws over a bridge, as it were,* [which leads] *to socialist transformations...*

... the establishment of [a "new type of democracy,"
transcending the bourgeois type but not yet socialist
itself] *would open the vista of a successful struggle
for the maturation* ['pererastanie', the same word
Lenin used to describe how the Bolshevik regime
supplanted Kerensky's in 1917] *of fundamental
democratic reforms into revolutionary socialist
transformations...*

*... The struggle for democracy, for profound democratic
reforms, is an intermediate step, leading to socialist
revolution.*

A Soviet historiography of Soviet studies of the
working-class movement notes (Rasputins, 1980: 139,
56-57; see also Aslapov, 1974) the essential difference
between these two tendencies (also represented by
Krasin, 1971 and Salychev, 1971). The united front was
based on cooperation only between communist and
socialist or social-democratic parties (the three-class
alliance) and asserted the main issue to be the solution
of "all the problems of [working-class] unity." For the
popular front, however, "the most important strategic
problem of the political avant-garde ... is to establish a
powerful and actively solidaristic union of all popular
forces" based on the broadest possible coalition of
social classes, including elements of the bourgeoisie
(the four-class alliance). Accordingly, Krasin (1974:
146) criticized a book by Salychev on the interwar
French Socialist Party for failing to take into account
the Communist International's work on "constructing a
new theoretical line ... in connection with rectifying an
underestimation of the role of democratic tasks in the
proletariat's class struggle." Salychev acknowledged

that the French Socialists in the 1930s did not solve "a series of questions of the theory of socialist revolution, especially the transitional form of power." But Krasin, writing in the very month when the French electorate was deciding between Giscard and Mitterrand as successor to Pompidou, emphasized Salychev's failure to develop this theme "with respect to its present-day significance," underlining that "the very same problem that then existed still undoubtedly exists today." The role of the Portuguese Communist Party in the Portuguese revolution directly posed this very same problem. It was the "theory of the three centers" of capitalism that made it again possible for the communists to consider incremental tactics towards the acquisition of power, and by electoral rather than violent means, in Western Europe.

More about structures

The foregoing section gave an example of how the existence of a cognitive hierarchy could affect value choice in Soviet foreign policy making. As time went on, the increasing organizational complexity of the Soviet foreign policy making Establishment created more possibilities for such hierarchically differentiated cognitions to recombine through multifarious channels of communication. In the early 1980s research on Soviet foreign policy making (Cutler, 1982a, 1982b, 1984b, 1985a, 1985b) led to the analytical specification of a Soviet press organization as a locus of roles within and around a press organ. This included both institutional and non-institutional roles, i.e., roles positionally inside

the press organ and outside it (cf. Rühl, 1969: 24-41).
The taxonomy of these roles was grouped into four sets:

1. The newspaper staff, most notably its editorial board
   and secretariat, plus its high-ranking journalists;

2. The *kruzhok* or "opinion circle" around the editorial
   board of the newspaper, which was however not
   part of the newspaper staff;

3. The particular institutional bureaucracy of which the
   newspaper was the "house organ", and;

4. Any "significant others" having ties to the
   newspaper who were important political figures in
   their own right, particularly members of the Central
   Committee and Politburo.

The institutional-group and interactionist theories'
structural postulates define which two elements of an
organization hold authority within the organization.
This specification of structural elements in interest
aggregation and articulation with reference to press
organizations is generally applicable, at a slightly
higher level of abstraction, to all post-totalitarian Soviet
organizations. The transformation of the Soviet press
system in the mid-1950s occurred against a background
where the third and fourth of these organizational
elements were the most significant determinants of
publicly articulated institutional policy preferences.
Figure 1 expresses this structure, which also represents
the logic of the institutional-group theory, which it
inherited from its parents the conflict-school theory and
interest-group approach.

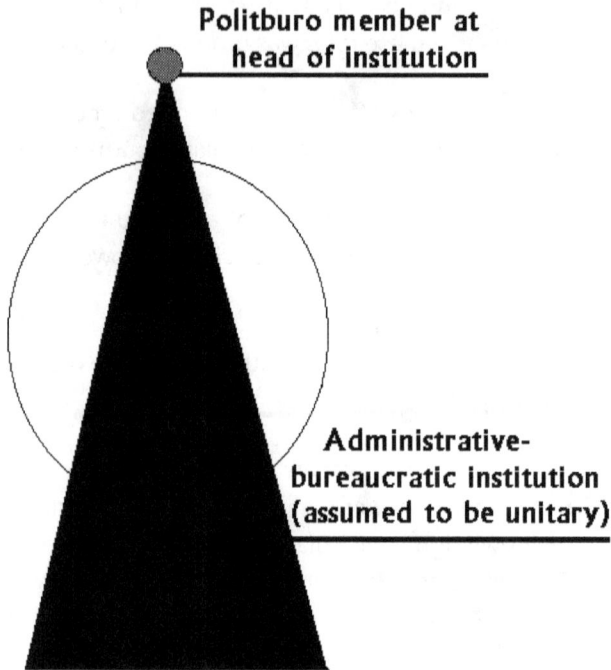

Figure 1 *Institutional-Group Theory*

Recruitment of personnel became more open after Stalin's death. A rough gauge of this is the fact that the membership of the journalists' union, formed in the mid-1950s, more than doubled between 1959 and 1971, from 23,000 to nearly 50,000. Over the same years, more than twenty programs were established in universities exclusively for the training of journalists (Hollander, 1972: 34-36). Routinization of education could in fact enhance control from above concerning personnel selection from a systemic perspective, but not so with reference to the individual editorial organizations. The expansion of recruitment and training programs in fact helped to mitigate any rigidification or standardization among recruited personnel. The wider variance in individual social and technical background among recruited personnel enhanced the diversity of views among the aggregate. Also the need to meet consistently increasing demand for journalists made selectivity itself more difficult to implement. At the same time, specialized training among journalists also increased, including in the foreign-affairs field (Vil'chek, 1973; Romanchuk, 1970).

Although the changes of the 1970s had an overall conservative effect, they did not reduce the transformations of the previous period but instead rationalized them. The new leaders who ousted Khrushchev convoked the Second Congress of the Union of Journalists in 1966 so as to give the media their marching orders, but did not really succeed in reducing the sources of available information. For example, toward the end of the 1970s, *Pravda* had nearly eight times as many of its own reporters overseas than it did a quarter of a century earlier; *Izvestiia*, which then had none, now acquired several dozen; *Trud* and *Komsomol'skaia pravda*, among others, also enjoyed the privilege (Tsukasov, 1975: 34-36). Organizations that had their own staff correspondents overseas did not have to rely so much on the standard digests circulated by Tass and Novosti. The situation was evolving to the point where the first and second above-mentioned organizational elements became the most significant determinants of an organization's policy preferences. Figure 2 expresses this structure; it also represents the logic of the complex-systems theory of Soviet foreign policy making.

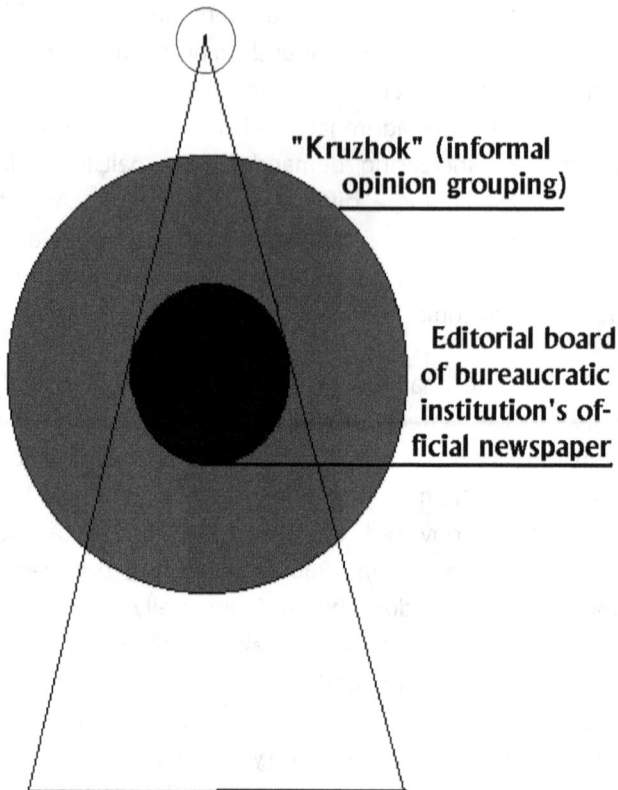

Figure 2 *Complex-System Theory*

By the end of the 1970s, individuals no longer made their careers inside a single propaganda organization. They had greater mobility among organizations, and therefore the control by individual organizations over the distribution of rewards was less rigorous. Indeed, there were more rewards and privileges to be distributed in general. Consequently, individuals were not as beholden to individual organizations and instead could circulate more freely among them (Dzirkals, Gustafson & Johnson, 1982; Lendvai, 1981).

This enhanced inter-organizational communication and further reinforced the tendency towards organizational fragmentation. The previous section of this monograph briefly discussed learning potential resulting from the recombination of cognitions. It is appropriate to ask whether there is a basis for inferring a learning potential from this organizational recombination.

Such a question of organizational learning can be dealt with here only most briefly. Organizational learning theory focuses on general domestic factors that influence any country's foreign policy. Three schools of organizational learning theory are particularly relevant. The first school is the new blood school. According to this school (Etheredge, 1981: 119), "resources are needed to continually hire skilled new people who bring fresh ideas or first-hand knowledge of what other people are doing." Beginning in early 1986, a wide-ranging personnel shake-up occurred throughout the Soviet Foreign Ministry, the Central Committee's International Department, and the entire foreign trade bureaucracy. The recruitment of more and younger personnel into the overlapping policy-making and propaganda-making systems in the mid-1980s made a "new blood" argument plausible for the Soviet Union.

The second school is communication flow theory, according to which "much innovation ... is embedded within the changing quality of communication" and "innovation rates are increased by particular patterns of communication and by networks of intra-institutional structures that create and support them" (Etheredge, 1981: 120). The cognitive component of the complex-systems theory implied that the quality of communication among those organizations was increasingly sophisticated. This characteristic would have facilitated the policy innovation that is associated with learning. Within the Soviet Foreign Ministry itself, new patterns of communication became evident which contrasted with the top-down autocratic manner in which Gromyko ran the ministry. For example, a "Collegium" was established in the ministry under Shevardnadze. It was a "formal consultative and decision making body" that met weekly and was composed of the deputy foreign ministers plus the heads of certain of the Ministry's departments elected to it. In a remarkable change from Gromyko's method of administration, any member of the Collegium could suggest any new idea on any subject discussed.

Finally, optimum norm theory argues that certain organizations, cultures, and climates—usually those that are "open, egalitarian, and problem-oriented organizations rather than status or career centered"—are especially conducive to organizational learning (Etheredge, 1981: 120-21). The organizational and bureaucratic culture of Russia and the Soviet Union would normally not be thought to have much learning potential under this last theory (Fainsod, 1963). However, Shevardnadze himself (1987: 32), in a speech to the personnel of the foreign ministry, insisted that "talent" be the criterion for advancement and promotion in the Soviet foreign service. If we consider not the norms motivating behavior within organizations but rather the change in those norms over time, then rapid changes under Gorbachev would have justified a plausible case for a certain degree of learning potential.

The structural-normative dynamics of complexity in the Soviet foreign policy making establishment

The last part of this monograph described structural and normative changes in the Soviet foreign policy making Establishment. This part describes their consequences for behavior. The most rigidly structured and constrained segment of the Soviet policy-making system was the one dealing with foreign policy. Many Soviet journalists working overseas for the Soviet press, covering international affairs in the main Moscow newspapers, were actually members of the Soviet foreign policy making Establishment. Within that Establishment, the least studied and most complex aspect is the intersection between the propaganda-making and policy-making systems. Overseas correspondents and editors at major Soviet newspapers were indeed some of the principal nerves of the Soviet body politic. Working journalists in the Soviet press not only collected and reported first-hand information on international affairs but also—some of them—influenced foreign policy by analyzing that information (Cutler, 1982a, 1982b). This occurred through links that journalists developed with party and ministry officials through the institutions of the media. The career of Georgii Ratiani exemplifies this overlap between policy-making and propaganda-making circles, between research and journalism in international affairs (Zhukov, 1980).

A large Western literature dealt with the participation in policy-making by the Academy of Sciences research institutes, but a high-quality Soviet literature on the working press was almost entirely ignored. This was due partly to the tendency for people in academia to be interested in each other, but it also reflected the absence of training and sensitivity on the part of most Western Sovietologists with respect to the standards of inference and demonstration that classically trained historians take for granted in interpreting documentary evidence. Although there were notable exceptions, Western Sovietologists who studied Soviet foreign policy, especially after Khrushchev's overthrow, tended, if they were not themselves émigrés, to be political scientists who did not care what standard operating procedures were used by the people who produced the data that they analyzed (i.e., by the Soviet editors and reporters who published the newspapers that they read).

Structural complexity in propaganda-making and policy-making

In the Brezhnev era, the overlapping systems of propaganda-making and policy-making in foreign affairs comprised the Communist Party's Central Committee (including its Politburo and Secretariat), the relevant State Committees and Ministries, the Soviet press, and the Academy of Sciences research institutes. Table 3 illustrates the differences between Steinbruner's (1974: 125-35) categories of "uncommitted" and "theoretical" thinkers, and helps to clarify their relationships. Different people use all three modes of thinking at different times, of course, but each mode characterizes certain definite roles.

Structural complexity in the policy-making hierarchy

In the policy making hierarchy, uncommitted thinkers were concentrated in the higher echelons, grooved thinkers in the lower, and theoretical thinkers in the middle. The low-level functionaries in the Ministry of Foreign Affairs were grooved policy-makers. The Ministry of Foreign Affairs applied the general foreign policy line to specific circumstances, consulting with different segments of the Central Committee and with specialists from the Academy of Sciences. The chief decision-makers in the country, including the Politburo and the Secretariat, and to some degree the Central Committee departments, were uncommitted policy-makers. Uncommitted thinkers use a more abstract framework and a more extended time frame than grooved thinkers. They deal with a greater range of problems and have a greater scope in which to address individual problems. The institute-based advisors and ministerial consultants were theoretical policy-makers (Eran, 1973; Glassman, 1968; Meissner, 1977; Petrov, 1973; Schapiro, 1975, 1977).

Specialists in international relations were active in advisory roles to foreign policy decision-makers well before 1953, and international-relations studies proliferated in the Soviet Union in the late 1950s and 1960s. Under Brezhnev, the Central Committee formulated the general line of Soviet foreign policy, but participation in the process was not necessarily restricted to the Central Committee. Every branch of the Central Committee participated in formulating Brezhnev's

	Uncommitted Thinking	**Theoretical Thinking**
Chief cognitive characteristic	Consistency and stability principles prevent overall integration foe divergent patterns of thought, each urged on the decision maker by a different "sponsor."	When highly generalized conceptions become established, they provide the mind with a basis for handling the uncertainty of the immediate decision problem.
Other characteristics	Due to the organizational setting, the reality principle forces a more abstract intellectual framework than for the grooved thinker. But abstraction is made difficult by uncertainty.	Beliefs are generally organized around a single transcendent value, inferentially related to specific objectives.
	The time frame is relatively more extended, with a greater range of problems and greater scope of individual problems.	Since thought processes are less dependent on incoming information to establish coherent beliefs, inconsistency mechanisms are widely employed to cope with it.
	There is oscillation among competing belief patterns, compromising stability somewhat, in favor of simplicity.	Likely to be found in small, closely knit groups which interact regularly over issues of common concern. This pattern of interaction provides social reinforcement.

Organizational condition in propaganda-making organizations	Within a particular information channel in an organizational unit, formal or informal, having a restricted scope of concern.	Levels where intersecting information channels carry relatively abstracted, aggregated information.
Examples in Soviet propaganda-making organizations	Operational propagandists, newspaper editorial board members, Political Observers, Special Correspondents	Propaganda planners and ideologists, members of social organizations
Organizational condition in policy-making organizations	Levels where intersecting information channels carry relatively abstracted, aggregated information.	Within a particular information channel in an organizational unit, formal or informal, having a restricted scope of concern.
Examples in Soviet policy-making organizations	Politburo, Central Committee departments, Central Committee Secretariat	Academy of Science institute researchers

Table 3 *Differences between "uncommitted" and "theoretical" thinkers, with examples from the complex system of the Soviet foreign policy making establishment. (Based on Cutler, 1981, drawn from Steinbruner, 1974; for further discussion, see Haas, 1991.)*

"Peace Program," announced at the Twenty-fourth Party Congress in 1971, and members of in the Academy of Sciences research institutes were also widely consulted (Eran, 1979: 18-29, 45-59; Day, 1981; Petrov, 1973; 824-27, 841-44; Garthoff, 1985: 42).

There were three ways in which individual "theoretical-thinking" specialists exerted influence. The first was to air ideas in institutional journals. Journal articles had greater immediacy than lengthy monographs and so carried more influence in policy circles. Ideas could also be aired at roundtables and conferences held in the institutes themselves. Officials from the Ministry of Foreign Affairs, cadres from the Central Committee, leading journalists, and other political observers participated in these exchanges of views (Löwenhardt, 1981: 87-89). A specialist's choice between the second and third instruments of influence depended upon the nature of the problem: an analyst could submit either a policy planning report or a situation report. In a policy planning report the Soviet analyst of international affairs did not express a policy preference. He only established the available options in general terms, laying out broad policy alternatives. Very much as in other countries, he could discuss the results he foresaw for each alternative, and the differing situations to which he sees them variously leading. This is frequently a means for expressing a covert policy preference (e.g., Inozemtsev, 1972, analyzed in Legvold, 1974: 17). Situation reports, on the other hand, were submitted on request by specialists at research institutes to one or another Foreign Ministry desk. Due to the often fast-moving nature of events, they were usually contracted on an *ad hoc* basis through informal and personal contacts (e.g., Vik. Kudriavtsev, 1974, analyzed in Cutler, 1985a: 66-67).

Structural complexity in the propaganda-making hierarchy

The allocation of roles in the propaganda making hierarchy differed from Steinbruner's generalizations. In particular, theoretical thinkers occupied the highest echelons and uncommitted thinkers occupied the intermediate ones. The "theoretical" thinkers were propaganda planners and ideologists, because they worked with highly generalized conceptions within a restricted scope of concern. They established the system's general propaganda strategy through regular meetings with high-echelon operational propagandists such as the editors-in-chief of the principal newspapers, to whom they issued directives. Some theoretical propaganda makers also constituted a subset of theoretical policy makers.

Directives issued by propaganda planners ("theoretical thinkers" in the propaganda apparatus) were aimed at individuals who had competence over a limited range of tasks, which they executed according to standard operational procedures. These operational personnel, such as low-level Novosti commentators or Tass correspondents stationed overseas, addressed uncomplicated problems that nearly always fell readily into a small number of types, and therefore were "grooved" thinkers. Their scope of responsibility was limited to expressing others' evaluations, such as may be received via propaganda directives, and to collecting and transmitting raw information for such evaluations. At this level of hierarchy in the information channel, there was the least scope for the expression of independent judgment and evaluation. Staff Correspondents stationed overseas for individual newspapers fell partly into the category of grooved thinkers, because collecting and transmitting raw information is an important part of their duties. Their authority to evaluate this information independently and to communicate those evaluations varied with the individual journalist, depending on his experience and his relations with the editorial board back home in Moscow .

Uncommitted thinkers used a more abstract framework and a more extended time frame than grooved thinkers. They dealt with a greater range of problems and had a greater scope in which to address individual problems. In the policy hierarchy these individuals occupied roles at the highest levels of organizations, but in the propaganda hierarchy they occupied intermediate roles between theoretical and grooved thinkers. The most important uncommitted thinkers occupied middle-ranking roles such as Political Observers, Special Correspondents, and editors of the international section in the newspapers' own offices. Journalists given the title "Special Correspondent," for example, were sent to investigate or to cover a specific event. A Special Correspondent sent overseas was familiar with the country he was going to and with Soviet policy toward that country. He "work[ed] with the confidence of the secretariat of the editorial staff" and was "entrusted with the organization of operational materials" (Pel't, 1980: 30). Anyone could be designated a Special Correspondent by the newspaper's editor-in-chief, usually subject to the approval of the "responsible secretary" of the newspaper. Often chosen from high in the bureaucratic organization, the Special Correspondent was an authoritative journalist who had the duty to evaluate the circumstances on which he was sent to report. Indeed, one of the complaints lodged against Soviet journalists in the field by their senior colleagues was that Special Correspondents did not do this often enough (Kudriavtsev, 1974). Journalists holding the title of Political Observer and Special Correspondent were able to articulate publicly their more personally held views. As a Soviet sociologist of the Soviet press observed, commentators holding the rank of Political Observer in particular had "the right not only ... to evaluate [international] events for which no official position has yet been established but also, taking the changing situation into account and based on independent analysis ..., to bring new nuances into the existing official position" (Popov, 1984: 34). On occasion, therefore, some uncommitted and theoretical propaganda makers also acted as, or expressed the views of, theoretical policy makers.

Normative complexity in propaganda-making and policy-making

Normative aspects of the intersection between propaganda-making and policy-making structures

Figure 3 depicts graphically the logical and political relationships among grooved, uncommitted, and theoretical thinkers in the overlapping propaganda-making and policy-making systems

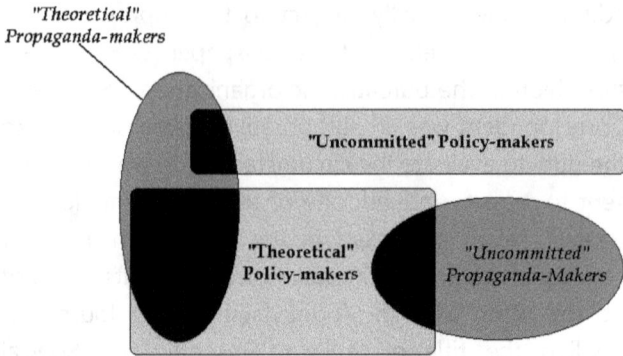

Figure 3 *Intersection Between Propaganda Making And Policy Making Roles In Soviet Foreign Policy Making Structures*

described in the last section. The insights it portrays are generated by the complex-system theory of Soviet foreign policy making, which also explains how feedback occurs from press analysis to policy making. An individual (such as a Special Correspondent) who believes that the model of the international environment implicit in the dominant policy tendency does not to coincide with newly received information, could write an article—or cause it to be written—that expressed an alternative view on the issue, thereby questioning the dominant view and generating either implicitly or explicitly a policy alternative. Indeed, unexpected international events motivated organizations and people in them to evaluate new information at hand, on the basis of their preferred models of the international environment.

Issues of this nature were hardly ever raised to the Politburo level. Normally, the standard operating procedures of the overlapping policy-making and propaganda-making organizations resolved spontaneous differences in the line to be taken. Although disputes could be resolved through an authoritative decision, the next unexpected international event would bring the overall "definition of the situation" into question again. Decisions on propaganda strategy thus not only expressed consensus about what to tell the people who read newspapers inside and outside the country, but also resolved very real differences of interpretation among press observers. In specific situations it is possible to identify specific journalists who performed this task of resolving such conflicts of interpretation (Cutler, 1982b, 1990). Usually, but not always, these were "theoretical thinkers" in the propaganda-making system who had other roles also in the policy-making system. Figure 3 depicts specifically such aspects of the intersection between propaganda-making and policy-making structures.

Figure 3 also expresses a set of organizational insights from the complex-system theory of Soviet foreign policy making that give a more nuanced portrayal of policy-making and propaganda-making than does institutional-group theory, which focused on institutions as carriers of ideas. It goes a step further even than the interactionist theory, which concentrated on the trans-institutional circulation of ideas even more than the individuals circulating them. The complex-system theory adds an emphasis on the level of abstraction of different ideas, and draws attention to how the recombination of these ideas is conditioned by the relationships among the individuals who carry them. Those relationships, in turn, are not only constrained by the norms of the organizations in which those individuals work, but moreover by the differentiation of their roles (set out in Table 3) in the overlapping propaganda-making and policy-making systems.

For example, the institutional-group theory would suppose that a difference of opinion between the Party's Central Committee and the government's Ministry of Foreign Affairs may have been reflected at some point in a difference between the Party's newspaper *Pravda* and the government's newspaper *Izvestiia*. They would, however, normally have tended to be close one another and to the main line of Soviet foreign policy. Also according to the institutional-group theory, *Pravda* and *Izvestiia* were the two principal press organs representing foreign policy views. Other press organs having a special role reflected institutional positions derived either from political function (e.g., *Krasnaia zvezda* and the military) or from specialized readership targeting. The complex-system theory adds that *Pravda* and *Izvestiia* had specific functions in the propaganda-making system that may have influenced their articulated policy preferences, and that other major press organs could also assume certain systemwide roles under special circumstances (Cutler, 1982b, 1985a, 1990). The complex-system theory also hypothesizes that institutional boundaries were more permeable than the institutional-group theory supposed.

The complex-system theory's definition of an organization (see Figure 2) further implies that differences between such institutions as the Party's Central Committee and the government's Ministry of Foreign Affairs were no longer necessarily even the most salient, because the institutions (and the people in them) were no longer "impermeable." This was especially true at the intersection of the Soviet policy-making and propaganda-making systems in foreign affairs. Since individual career paths were no longer made within a single institution after Stalin, it became easier for people in foreign policy making organizations and people in the propaganda making organizations to coalesce into trans-institutional "tendencies of articulation" (Griffiths, 1971). In fact, the institutional-group theory suggested that the Party's Central Committee should have been more orthodox with respect to the strategy of West European communist parties, and the government's Ministry of Foreign Affairs more heterodox. However, exactly the opposite was the case, because the Central Committee's greater contact with (and hence exposure to) the West European communists drew it to the "Right," while the ossifying conservatism in the Ministry under Gromyko, which proved so hard for Shevardnadze to change, drew it, on the other hand, to the "Left."

Normative consequences of the intersection between propaganda-making and policy-making structures

What does all this imply for foreign policy learning? According to Deutsch (1963: 96, 210, 222), foreign policy learning may be cognitively manifested either through the transformation of goals held at the outset into goals not previously conceived, or through the choice of preexisting alternative goals over other goals originally held; Haas (1991) reserves the term "learning" for goal transformation, as distinct from "adaptation." The transformation of goals held at the outset into goals not previously conceived is rare and cannot be programmed. An example is the Soviet reconciliation with the absence of revolution in the West during the 1920s and early 1930s. The failure of proletarian revolutions during and after World War One resulted in removing the ("Left") strategic goal of universal revolution to the level of ideological prescription. Its direct influence on tactics correspondingly diminished. The doctrine of "socialism in one country" and its associated foreign policy confirmed and legitimated this cognitive rearrangement. The late 1950s and early 1960s were a period when altered Soviet perceptions led to attitude change about world politics in general and other international actors in particular, through the choice of preexisting alternative goals (Zimmerman, 1969). Griffiths (1972: 472; see also Griffiths, 1981, 1984, 1991) concluded that during that time "the [Soviet] need to learn new patterns of international behavior determined what was learned about the United States." The resources invested in attaining incremental ("Right") minimum goals could imbue these goals with additional value, to the point where they eventually supplanted the original ("Left") maximum goal, relegating it to the more abstract reaches of the cognitive hierarchy. This removed it from direct influence on day-to-day policy. The mid-1950s and early 1970s were also periods during which broad goal transformations did occur in Soviet foreign policy behavior.

Recall that *doctrine* is the immutable historical teleology, and *ideology* is the set of linguistic categories that expresses the doctrine. Thus the dispute over whether the world capitalist system had one or three centers was a question of ideology. When the categories of ideology concatenate to define what really exists, they define the set of possible futures because what is inexpressible in their language becomes impossible.

Specific conceivable futures imply goal-oriented lines of action to their attainment, and each goal-oriented line of action is a *strategy*. So when Pevzner, a Soviet advocate of the "theory of the three centers" of world capitalism argued (reported in Graboski & Nowak, 1969: 192) in favor of that ideological revision because studying "the conflicts ... between 'Europeanists' and the supporters of an 'Atlanticist' policy, ... [enables us to oppose more effectively those] forces that try to influence West European integration an antisocialist direction," this was a justification on the level of strategic efficiency. *Tactics* operationalize strategy at the level of actions, intentions to act, and preferences for actions to be taken, with reference to concrete issues in particular policy fields.

That being so, as explained above, two related strategic and tactical issues pertinent to Soviet policy making toward Western Europe in the first half of the 1970s distinguished the orthodox and heterodox models of the environment from one another. The first issue was how seriously to take the EEC. This concerned the consolidation of West European economic integration and especially its acquisition of autonomous political qualities. The second issue concerned prescription of the role that West European communist parties should play in West European politics. This was about the desirability of communist participation in West European governments, especially the feasibility of their using "bourgeois" parliamentary democracy to attain "socialism." As already explained, different Soviet attitudes in the early 1970s, toward "Europeanists" and "Atlanticists" in the West, reflected opposing Soviet views of the effects on NATO of expanded EEC integration.

On the Left (i.e., according to the orthodox model of the international environment), the expansion of West European integration was seen as a façade for increasing American influence on the EEC, because Great Britain, then on the verge of EEC membership, was thought to be the driving force behind moves by NATO's "Eurogroup" towards West European military integration. Analysts adopting this tendency naturally thought that the 1974 Cyprus conflict had erupted at NATO's instigation. For not only were Greece and Turkey both NATO members: moreover, under international treaties they were supposed to "guarantee" the island-state's independence; and the third signatory-guarantor of the independence accord was Cyprus's former colonial master, Great Britain! These ideas supposed that the U.S. could still get Greece, Turkey, Britain, and other West European governments to do its bidding. This view therefore ran counter to the essence of the "theory of the three centers."

When the "theory of the three centers" was introduced, the *doctrine* remained unchanged, but the *ideology* evolved; and this had real implications for the tactics of the communist movement and the Soviet state. The ideological theory of the three centers meant that West European political integration had to be taken seriously but did not inevitably lead to military integration. A corollary of this was that a West European communist party might come to power in its own country without cascading communist revolutions throughout the whole of Western Europe, let alone the whole world capitalist system (Cutler, 1980). In Marxist-Leninist language, that meant coming to power without installing a dictatorship of the proletariat. The *strategy* of the "democratic alternative" was therefore introduced. It advocated joint political and social action by communists with other left-wing forces, and proponents of the "theory of the three centers" explicitly invoked it.

Therefore, on the Right, proponents of the heterodox model of the international environment contended that the Cyprus conflict in 1974 resulted from long-standing Greco-Turkish rivalry having little to do with NATO. Also according to this viewpoint, the expansion of the EEC would help to realize the "democratic alternative" by mobilizing politically the working classes and intermediate social strata. This led in turn to *tactical* changes in the actual practice of the world communist movement (Alekseeva, 1971). In early 1974 French President Pompidou died suddenly, and the French Communists implemented a "Common Program" with other leftwing parties that included support for the socialist Mitterrand, even in the face of embarrassing Soviet support for the centrist Giscard (who had Gaullist support). Soon thereafter came the bloodless April 1974 revolution in Portugal, overthrowing the decades-old fascist dictatorship there, and leaving the once-underground Portuguese Communists in a politically strong position, but allied with other antifascist parties, including not only the socialists but middle-of-the-road bourgeois parties that wanted only to establish a multiparty parliamentary system on the West European model (Cutler, 1982b, 1982c; compare Révész, 1975, 1976; Troianskii, 1980; Wettig, 1975).

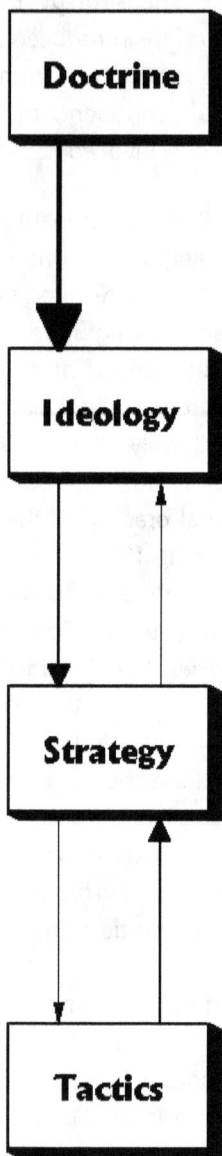

Figure 4 *Relative strength of inter-level constraints in the cognitive hierarchy*

Figure 4 illustrates the nature of constraints upon attitude change and policy innovation within the hierarchically organized belief system characterizing Soviet foreign policy making. Tactical modifications could aggregate up to the level of altering strategy, but it was much more difficult for modifications of strategy to alter the ideology. In general, ideological change required prior doctrinal legitimation; and doctrinal change could apparently occur only with the rise to power of a new General Secretary of the Party, and not necessarily even then. The doctrine constrained acceptable ideology, whether on the Left or on the Right.

Empirical work reveals that a further nuance is required. Organizational learning is affected by the character of the external environment. The environment should be characterized not only by objectively determined qualities (in particular, the length of a situation's duration) but also by subjective qualities perceived by the organizational actor (in particular, the amount of tension experienced). The interaction of these two dimensions may be interpreted as a degree of stress, where short-duration and low-tension situations are lowest in stress, and long-duration and high-tension situations are highest. Research reveals (Cutler, 1990) that:

1. In general the least stressful situations (those characterized by short duration and low tension) engender incremental responses to the environment that are of a mainly tactical nature and concern the relation between tactics and strategy.

2. Long-duration, low-tension situations principally invoke relations between strategy and ideology, leading to the possibility of revision at the ideological level in the medium or long term.

3. Still more stressful situations, typically characterized by short duration and high tension, instantiate relations between doctrine and ideology.

4. The most stressful situations (long duration and high tension) tend to freeze and institutionalize existing relations among all levels in the cognitive hierarchy.

This last type of response results from the need to exercise some sort of control, however limited, over the environment. When the environment has so high a degree of uncertainty, action without specific reference to the environment becomes the last resort. A good example is the nature of U.S.-Soviet relations under Andropov. At a time when nuclear arms negotiations between the superpowers were at a standstill in Europe, and when American intransigence on deployment of the Strategic Defense Initiative (SDI) was sharpest, Soviet diplomacy moved into a straitjacket. It ruptured arms-reduction negotiations and undertook an international propaganda campaign of remarkable venom directed at President Reagan personally, the vitriol literally unmatched in duration and intensity since Stalin's death. Thus a prolonged situation of high tension resulted in a unilateral Soviet decision to take a single course of action and stick to it simply in order not to feel entirely at the mercy of events. From the standpoint of the actor, this is a defensive institutionalization of an offensive policy. However, the danger is that it limits if not altogether abolishes the possibility for learning.

So, all other things being equal, this situation is worst when the environment is characterized by a continuing presence of high-tension, short-duration situations. Of course, this is most frequent in contemporary corporate management. A glance at Figure 4 shows why it can be deadly. It suggests why some organizations have more difficulty than others dealing with stressful situations. In tandem with Table 3, it illustrates also, for example, how ideologists and high-level propagandists, whose career advancement has been due to personal investments in doctrine and ideology, make it hard for any organization to motivate change. Naturally, the greater their influence among the highest decision-makers, the still harder will be such change.

Normative constraint and learning from the environment

The Soviet Union always paid special attention to its public diplomacy and sought to attune its propaganda making to foreign audiences. In trying to influence the international audiences of this public foreign policy, Soviet policy makers increasingly took them into account not only in propaganda-making but also in policy-making. The mutual cooptation of propaganda specialists into policy-making and policy specialists into propaganda-making, facilitated adaptation and change and even learning in Soviet foreign policy, precisely because the Soviet Union was always attuned to propaganda-making to foreign audiences and always paid special attention to its public diplomacy. The reason for this is that Soviet foreign policy was informed by a doctrine called Marxism-Leninism, where categories of sociological analysis had a long and traditional place of pride.

Marxism was originally not a stultified economic theory but part of a very dynamic and creative social-theory movement in the mid-nineteenth century that built on categories from German idealist philosophy because sociology had not yet been invented (Marcuse, 1960). The world communist movement was a movement of national parties and national publics, of real people rooted in historical time and culture, in geographical space, and in national economies composed of actual commodity flows rather than some idealized "economic system." It was emphatically not an aggregation of the ideal *homo economicus*, this Western delusion that misguided Western economic policy toward the post-Soviet economies with effects so deleterious to those societies, that even its strongest exponents have now come publicly to admit their culpability for the policy errors that their own ideological blindness has imposed upon those countries (Greenspan, 1997).

Soviet public diplomacy goes back to the 1918 Treaty of Brest-Litovsk, where the conquering Germans were astounded at the Bolshevik insistence that Russians be allowed to fraternize with German soldiers. Likewise, Soviet Foreign Minister Chicherin's 1922 speech at Genoa was intended as an appeal over the heads of Western governments to their publics. Under Lenin, however, the Soviet press was not developed enough really to be a tool of this public diplomacy. Under Stalin, the Soviet press was no longer really an instrument of public diplomacy apart from its basis propaganda function, but starting under Khrushchev there was a broader opening to the world as early as the mid-1950s that paralleled the increasing importance of communications and public diplomacy in the world at large (Roth, 1980). The 1959 Congress of Journalists as well as the establishment of the specialized foreign-propaganda press organization Novosti in 1961 were part and parcel of that development, which in the 1970s was integrated into doctrine which identified the Soviet press and international information policy as an instrument of the "ideological struggle."

For other political reasons, at the same time, each new leader also put together a foreign policy program: e.g., Khrushchev's "peaceful coexistence" and "zone of peace"; Brezhnev's "Peace Program," "détente," and "relaxation of international tensions"; Gorbachev's "mutual security," and "new political thinking." Getting the domestic policy into line also included a restructuring of the domestic propaganda apparatus, following which a restructuring of the foreign propaganda apparatus typically occurred. It was always the case that careers in foreign propaganda making and foreign policy making overlapped in the Soviet Union. Under Gorbachev, more and more people moved between policy-making and propaganda-making organizations. The difference with Gorbachev is that he recognized the real importance of public diplomacy in the modern age (Koschwitz, 1979). For instance, Evgenii Primakov (1987), the Russian foreign minister who was a close advisor to Gorbachev, publicly acknowledged that the Soviet decision in February 1987 to separate the issue of medium-range (INF) missiles from the Reykjavik package of arms-reduction proposals was greatly influenced by the position of West European countries and by feedback from an international conference of nongovernmental organizations called the Moscow International Peace Forum.

Also in 1987, Gorbachev reversed a decision not to visit Washington after he discovered that this did not play well in Western Europe. Gorbachev's initiative for a "Common European Home," while based on the need for openness in Soviet foreign policy in order to facilitate the domestic policies of perestroika, also had an important public-diplomacy aspect that transcended mere public relations. Gorbachev sought to define a new line called "new political thinking," partly in order better to discern those who were for and those who were against him in the domestic arena. It was important for him to clarify this new political thinking, and that helped to create a more unified political coalition. "New political thinking" was itself an instrument of foreign policy under conditions of glasnost, and public diplomacy was part of its external aspect.

Every Soviet leadership succession initiated a cycle that meant consolidating a new general political line having foreign and domestic policy components. When leadership successions occurred in the Soviet Union, the successful successor reformed or otherwise reorganized the press system because of the purely political necessity of putting the propaganda apparatus into order and giving them their marching orders. Thus international affairs commentary in the Soviet press was one of the last topics affected by glasnost in the Soviet press. As early as mid-1983, two years before Gorbachev, the Central Committee held a plenary meeting on "counterpropaganda" and announced the need to include more direct and factual and firsthand information in reports on the West, and on the U.S. in particular. It called for such reports to be counterbalanced by commentaries reflecting the Party line. This represented an official endorsement of the classification of Soviet press commentaries on international events into three categories: "propagandistic" ones, which recount positive Soviet foreign policy steps, "critico-propagandistic" ones, which comment on the foreign relations of capitalist countries; and "counter-propagandistic" ones, which are designed to combat anti-Soviet tendencies that characterize foreign media (Popov, 1981; cf. Beglov, 1980). Analogous categories are not unknown in the public relations strategies of certain corporations.

Izvestiia's Aleksandr Bovin had always been among the more heterodox Soviet press commentators on foreign affairs. One of his articles shows unequivocally how publication can be a step in the direction group advocacy of an alternation in the image of the international environment. In early 1987 he criticized the late-1970s' deployment of SS-20 missiles in Eastern Europe was one of the first instances of a journalist overtly criticizing any aspect of post-1945 Soviet foreign policy before such a criticism has been authoritatively made by leading political authorities and thus sanctioned. This launched a lengthy discussion in the scientific journal of the leading international-relations institute in Moscow, which came to the conclusion that the deployment was a mistake but not as big a one as Bovin had thought. Interestingly,

Bovin's criticism of the Soviet stand at the Reykjavik talks with President Reagan was not endorsed (Bovin, 1988; Karaganov, 1988; Semeiko, 1988; Sturua, 1988) by other writers. At about this same time, Soviet scholars and journalists adopted the very terms "image of Self" and "image of Other" from Western social science, and complaints began appearing in the Soviet press that clichés about the West and primitive stereotypes of the "concept of enemy" had not disappeared (Znamenskaia & Smagin, 1988). This suggests how difficult it was to implement the resolutions of the 1983 plenary session on counterpropaganda (cf. Zak, 1976). This suggests that Gorbachev's *doctrine* of "new political thinking" in international relations had some support in the USSR among the "uncommitted" propaganda-makers on the level of *strategy* but still ran up against resistance from the *ideologically* motivated "theoretical" propaganda-makers, who managed the propaganda institutions and exerted significant influence in political decision-making circles (see Figures 3 and 4).

What it all means for complex organizations today

If ideologues are disease vectors, is corporate culture a "fatal distraction"?

In a nutshell, the organizations in the Soviet foreign policy making Establishment auto-complexified in response to the increasingly complex global environment. Gorbachev finally accelerated changes in Soviet foreign policy doctrine, but the complex multiplication of political resources and incentive structures in Soviet society had already made that society effectively part of the global environment external to the Soviet political system. Consequently, the constituent parts of the USSR self-organized their own foreign policies independent of Moscow (Matlock, 1995: 409-18). In the end, what impeded Soviet foreign policy adaptation was doctrinal constraint upon organizational cognition and interest-articulation. The Soviet system imploded and collapsed due to accumulated inertia, under the explosive force of too much change too fast. People were not allowed to say they saw things that, according to the doctrine, were not permitted to exist; even the ideology could not be modified enough (Remington, 1985b). That is why, for example, to discuss the nationalities questions in the late 1980s, it was necessary in the USSR to invent a whole new language with new analytical terms having meanings that the previous rules of discourse did not permit to be recognized.

Under Khrushchev, the Party's Central Committee sent fact-finding propaganda groups into the regions to find out exactly what was happening on the ground (Hoffmann, 1968). (Recall that we are talking about the late 1950s, barely a decade after the defeat of Hitler, so communications and system feedback are actually a big problem.) At first, these fact-finding missions talked only with the local officials and administrators. Only a few years later, in the early 1960s, did Moscow realize that this was insufficient and sent such groups went back out into the field to see what was really happening. (Recall that at this time, the most elementary techniques of survey research were not yet fully developed in Western social science, and that IBM machine cards were not widely available in the West, and that the technology of inputting programs on punched paper tape had not yet been innovated.) When Khrushchev was overthrown in 1964, his successors redirected these developments in organizational feedback with the results that Soviet society eventually became part of the political system's environment rather than part of the system itself. For corporations, things get still worse. The corporate analogue of this "Soviet society" includes such elements as the customer base, suppliers, advertisers, and even management consultants. The failure of the central control mechanism of the corporation as system, to receive and properly interpret feedback, means of course that the system is increasingly fallible and unable to respond to demands emanating therefrom. That is how the "society" sector becomes part of the environment, no longer an internal environment of the corporation, but instead radically divorced from communications with the "elite."

Another shortcoming was that the Soviet Union denied scarce political resources to those articulating the new ideas. For example, beginning with studies in the Soviet Union of the "green movement" in the West and the impulse for environmental studies given by international institutions (see Rasputins, 1980: 36-46), Soviet political scientists (who never had a separate discipline of "political science" in the universities or the Academy of Sciences) took the initiative in the late 1970s and early 1980s, to establish a separate academic discipline called "global studies" (*globalistika*). They failed in large part because their approach would have called for denying that Third World poverty was exclusively the legacy of colonial imperialism, and because it would have implied the need for fundamental cooperation between the socialist and capitalist world-systems to alleviate such problems (Cutler, 1984a; Valkenier, 1983). This failure meant that they would not have their own institute, which would not have its own journal, which would not legitimate their point of view. The significance of such resources in the Soviet context is evident if one reflects that a large proportion of academics and researchers whom Gorbachev sponsored and coopted in the beginning were from the Siberian branch of the Academy of Sciences in Akademgorodok. This set of institutions had been established by Khrushchev a third of a century earlier and had developed its reformist and critical ideas in relative isolation, i.e., without excessive interference from Moscow. They constituted a formidable human and bureaucratic resource by the time Gorbachev appeared on the scene.

Self-image as a brake upon environment-driven learning

The complex-system theory implies that the sub-institutions may be as affected and driven by the international environment as by their own domestic political environment. The degree of their earlier subordination under Khrushchev and Brezhnev to the political system and to the political elite led to the system's failure to adapt and change its ideology and doctrine to permit efficient strategic changes. That failure created the situation where things exploded. In fact we know that Gorbachev's most striking doctrinal innovations in foreign policy were of a nature that only the supreme political leader could make. Not even Khrushchev was entirely successful in his attempt to do the same, since the ideological constraints on strategy made it very difficult for the ideology itself to be revised on the basis of aggregated tactical lessons translated through strategy. Yet these ideological constraints may well interfere with the continued viability of the system as a whole.

If we look at the Brezhnev era, between Khrushchev and Gorbachev from roughly the mid-1960s to the mid-1980s, it was not quite that case that organizations in the Soviet foreign policy making Establishment could not say what they thought because of official doctrinal constraint, although this was of course to some extent true. The situation was a bit more nuanced. More precisely, it was the case that those to whom they might articulate these views had incentive structures that made them deaf. The danger came (see Figure 3) from the "theoretical propaganda-makers" stuck in old ways of thinking and who imposed these upon "uncommitted propaganda-makers" as well as upon the "uncommitted policy-makers" whom they advised. In a corporate environment, the doctrine is often called "corporate culture," and the "theoretical propaganda-makers" often themselves are "uncommitted policy-makers" because they may be, for example, board members inherited from earlier organizational developments within the corporation. These individuals have an established way of doing things, and in the worst cases are straitjacketed by the corporation's founder by being his (or her) biological children.

The interactionist theory suggested that learning in Soviet foreign policy was driven by the international environment, and that the effectiveness of that learning was manifested in changes in the images of that environment and of other actors in it. The complex-system theory goes further, to imply that these predispositions are learned only on a strategic level, but also that they are learned and influenced by the international environment with respect to specific policy issues of tactics on a cognitive-map level. An aggregate of things that are learned on the tactical level may "bubble up" to the strategic level; on rare occasions, an aggregate of things learned on the strategic level can bubble up to the ideological level. This occurred in the 1920s, when the first historical precursor of the "popular front" appeared in Soviet ideology under conditions of a high-stress international environment (Degras, 1967). If in the 1970s a high-stress environment militates against learning, what facilitated apparent learning at the earlier time? The answer seems to lie in the capabilities of

Soviet power, and in the leadership's assessments of those capabilities; for in the earlier period, the Bolshevik position in world politics was precarious.

We seem justified in concluding that a country (or a corporation) is much more likely to learn necessary survival behavior under conditions of institutional weakness in relation to the environment than under conditions of strength in high-stress situations. It would seem that self-awareness of one's own resource weakness automatically decreases the tension experienced: one spends less energy trying to rigidify useless or nonexistent internal structures. This in turn facilitates learning on the ideological level, even without excluding bubble-up to the doctrinal level. In practical terms, that means that it becomes permissible to redefine the entire organizational mission. Therefore the tentative conclusions enumerated in the middle section of the previous part of this chapter are appropriate only for situations where the organizational actor judges itself to have sufficient resources to withstand "psychological assaults" from the environment. When such resources are available, an organization is more likely to attribute its own real failures to aspects of the environment. It may then misallocate those resources in seeking to control that environment rather than in promoting necessary organizational change.

In the Soviet context, this insight complements Naylor's (1988) argument; and it is a truism in Russian studies, that domestic reforms in the Russian Empire and the Soviet Union historically tended to follow upon defeats in international politics. The ultimate lesson for governments may be that their countries do not have a function so transcendent in international affairs as to map into null-space their neighbors far and near. Germany required two world wars to learn this lesson, and it seems that Russia is slowly reconciling herself to it after centuries being the regional hegemon. The consistent overestimation of one's own resources leads only to over-extension, which in the best case produces defeat and in the worst case disintegrative collapse. To find analogies to this process in the lives and deaths of contemporary corporations is left as an exercise to the reader.

Coming to grips with what cannot be grasped

One solution is for organizational diagrams to cease being a management tool, and for merger and acquisition to predominate. Yet merger and acquisition in the corporate world today are barely survival tools by which corporations and their components are driven: driven, moreover, not necessarily by any internal logic about the environment but by the environment itself unreflectively. Prominent CEOs frequently face, with respect to their subsidiaries, much the same problem Gorbachev faced with the republics. However, if merger-and-acquisition becomes the dominant ideology, then corporate executives become less able to respond optimally to new problems that require solutions other than merger and acquisition. In the history of international politics, the examples of Napoleon and Hitler suggest themselves as analogies.

Consider Lithuania, for example, as a division of another company that was acquired by the "USSR, Inc.," under Stalin's stewardship. The military occupation of Lithuania was for Stalin a rational decision. Yet it was after a visit to Lithuania, already self-organizing to take itself out of the Soviet Union, that Gorbachev began talking about "socialist pluralism" and began planning to institute changes in the Party's rules that legitimized factions and erased its "leading role" from the long-standing doctrine. This would probably have happened in any event, given Gorbachev's intents for reform; but in the absence of the annexation of the Baltic states nearly a half-century later by Stalin, it could well have happened differently, with different results. There is an argument to be made, that Stalin's annexation of the Baltic states strongly contributed to the ultimate disintegration of the Soviet Union. In political terms, then, Lithuania was a "loss-making center" that should have been divested. But it was again the doctrinaire ideologues who created the political weight making such change impossible until it was too late.

Even in the late 1980s, if done properly this would not necessarily have meant the break-up of the Soviet Union. Gorbachev tried "downsizing" the USSR by drafting a new treaty to replace the 1922 agreement that formed the USSR—a "new Union Treaty"—and there were as many as nine republics that would have signed onto it as late as summer 1991. By rejecting the so-called "Shatalin plan" for economic reform, however, Gorbachev denied the leaders of those republics any incentive to stay in a reformed USSR, which he proposed to call the Union of Sovereign States (Matlock, 1995: 498-517). With no centrally generated rationale for remaining part of the parent organization, these divisions of "USSR, Inc.," driven by the international environment including their own electorates (which had become part of the Soviet Union's international environment while the Kremlin was not looking), spun themselves off. They found the international environment to be extremely receptive to this self-initiated antitrust action.

On the one hand, corporate subsidiaries have to have a fair amount of freedom to respond to the environment, both for their own efficiency and to guarantee the survival of the overall system. If that does not happen, then a corps of ideologues can form in the corporate headquarters that has a vested interest in the continuation of a dysfunctional corporate culture. Such a corporate culture is like a "field," invisible on organizational charts but exerting an influence on perceptions and decisions much as Marxism-Leninism did in the Soviet Union. In the end, either key engineers may leave and began their own spin-off start-up firms, or entire divisions may become unprofitable and subject to hostile takeovers and buyouts: all because the corporate ideologues have exerted such a pull that, like Khrushchev, even a "reformist" CEO may be unable to escape their influence.

On the other hand, if the subordinate managers are empowered to respond as necessary to their immediate business environment, relatively free of central direction, then there is no guarantee *a priori* that they will not decentralize themselves out of the system. That is what some of the Soviet republics began doing even in late

1980s, before the entire Soviet system came crashing down after the failed coup attempt in August 1991. The difficult question for managers at the highest level is how to allow the subsidiaries this freedom to survive and be viable, while at the same time providing them incentives to stay within the overall corporate institution, yet without imposing upon them constraints that sub-optimize their survivability.

There may be a trade-off between survivability within established structures on the one hand and, on the other hand, performance. Maximization of performance may sometimes be impossible under conditions of optimized survivability within a structure of subordination. Anyone who has ever worked for anyone else probably knows this from personal experience. There is no reason why organizational performance should be any different in this respect. Indeed, the more complex any environment becomes, the more likely this may be and the more frequently the contradictions may appear. The short duration of decisional situations and the experience of high tension in the contemporary business environment then combine to increase the stress on decision-makers. As pointed out above, all other things being equal, this decreases their ability to learn from the environment against which their very success or failure will be judged. Such a condition today seems characteristic if not fundamental.

The only remaining task is to determine how to prevent "complexity" and "complexity theory" from becoming another ideology, and management consultants from becoming doctrinaire ideologues with their own vested interests. Marx decried the fact that goods produced under capitalism became invested by the market with powers transcending their economic value: this was the famous "fetishism of commodities." However, that danger is at least as great with regard to commodities in the marketplace of ideas; and applied research on organizations exists in such a market. How to avoid the "fetishism of complexities" deserves more detailed treatment than is possible in these concluding remarks, however, and this topic is reserved for another work.

References

Abelson, Robert P. 1973. "The Structure of Belief Systems."
Pp. 276-339 in *Computer Models of Thought and
Language*, edited by Roger C. Schank and Kenneth Mark
Colby. San Francisco, Calif.: W. H. Freeman.

Adomeit, Hannes. 1979. "Soviet Perceptions of Western
European Integration: Ideological Distortion or Realistic
Assessment?" *Millennium* **8** (Spring): 1-24.

Alekseeva, N. V. 1971. "O nekotorykh problemakh strategii
i taktiki FKP." *Vestnik Moskovskogo universiteta*, ser. 13,
Teoriia nauchnogo kommunizma, no. 5 (September-
October): 41-50.

Anikin, A. 1971. "Valiutnyi krizis kapitalizma: prichiny i
posledstviia." *Kommunist*, no. 10 (July): 91-102.

Anikin, A. V. 1973. "Zoloto-valiutnyi standart: problemy i
protivorechiia." *Den'gi i kredit*, no. 3 (March): 77-86.

Aslapov, L. 1974. "Nekotorye problemy internatsional'nogo
sotrudnichestva kommunistov i sotsial-demokratov."
Pp. 34-53 in *Opyt i perspektivy sovmestnykh deistvii
kommunistov i sotsialistov: Materialy nauchnoi
konferentsii, sostoiavsheisia v AON pri TsK KPSS, 18-19
aprelia 1973 goda*, edited by I. M. Krivoguz (resp. ed.), K.
M. Obyden, and V. T. Nesterova. Moscow: (AON pri TsK
KPSS).

Aspaturian, Vernon V. 1966. "Internal Politics and Foreign
Policy in the Soviet System." Pp. 212-87 in *Approaches to
Comparative and International Politics*, edited by R. Barry

Farrell. Evanston, Ill.: Northwestern University Press. Reprinted at pp. 491-551 in Aspaturian, 1971.

Aspaturian, Vernon V. 1971. *Process and Power in Soviet Foreign Policy*. Boston, Mass.: Little, Brown.

Aspaturian, Vernon V. 1972. "The Soviet Military-Industrial Complex—Does It Exist?" *Journal of International Affairs* **26**, no. 1: 1-28.

Axelrod, Robert. 1972. *Framework for a General Theory of Cognition and Choice*. Berkeley, Calif.: University of California, Institute of International Studies.

Axelrod, Robert. 1973. "Schema Theory: An Information Processing Model of Perception and Cognition." *American Political Science Review* **67**, no. 4 (December 1973): 1248-66.

Axelrod, Robert (ed.). 1976. *Structure of Decision*. Princeton, N.J.: Princeton University Press.

Beqlov, S. I. 1980. *Vneshnepoliticheskaia propaganda: Ocherk teorii i praktiki*. Moscow: Vysshaia shkola.

Bezymensky, L. 1972. "Anti-Europe." New Times, no. 17 (April): 22-26.

Bialer, Seweryn (ed.). 1981. *The Domestic Context of Soviet Foreign Policy*. Boulder, Colo.: Westview Press.

Blackwell, Robert E., Jr. 1972. "Elite Recruitment and Functional Change: An Analysis of the Soviet Obkom Elite, 1950-1968." *Journal of Politics* **34**, no. 1 (February): 124-52.

Blackwell, Robert E., Jr. 1973. "The Soviet Political Elite: Alternative Recruitment Policies at the Obkom Level; An Empirical Analysis." *Comparative Politics* **6**, no. 1 (October): 99-121.

Bolotin, B., and Kudrov, B. 1972. "Tri tsentra v mirovom kapitalizme (Opyt mezhdunarodnykh ekonomicheskikh sopostavlenii)." *Mirovaia ekonomika i mezhdunarodnye otnosheniia*, no. 3 (March): 96-108.

Bovin, A. E. 1988. "Inye varianty." *Mirovaia ekonomika i mezhdunarodnye otnosheniia*, no. 12 (December): 28-32.

Breslauer, George W., and Tetlock, Philip E. (eds.). 1991. *Learning in U.S. and Soviet Foreign Policy.* Boulder, Colo.: Westview.

Cohen, Bernard C. 1970. *The Press and Foreign Policy.* Princeton, N.J.: Princeton University Press.

Converse, Philip E. 1964. "The Nature of Belief Systems among Mass Publics." Pp. 206-61 in *Ideology and Discontent*, edited by David Apter. New York: Free Press.

Cutler, Robert M. 1980. "The View from the Urals: West European Integration in Soviet Perspective and Policy." Pp. 80-119 in *Western Europe's Global Reach: Regional Cooperation and Worldwide Aspirations*, edited by Werner J. Feld. New York: Pergamon Press.

Cutler, Robert M. 1981. "Decision Making and International Relations: The Cybernetic Theory Reconsidered," *Michigan Journal of Political Science* **1**, no. 2 (Fall 1981): 57-63.

Cutler, Robert M. 1982a. "The Formation of Soviet Foreign
    Policy: Organizational and Cognitive Perspectives," *World
    Politics* **34**, no. 3 (April): 418-436.

Cutler, Robert M. 1982b. "Soviet Debates over Foreign Policy
    towards Western Europe: Four Case Studies, 1971-1975."
    Ph.D. diss., University of Michigan.

Cutler, Robert M.   1982c. "Unifying Cognitive-Map and
    Operational-Code Approaches: A Theoretical Framework
    and an Empirical Example." Pp. 91-121 *Cognitive
    Dynamics and International Politics*, edited by Christer
    Jönsson. London: Frances Pinter.

Cutler, Robert M. 1984a. "Economic Issues in East-South
    Relations." *Problems of Communism* **33**, no. 4
    (July-August): 73-80.

Cutler, Robert M. 1984b. "Organizational Processes in Soviet
    Foreign Policy Making." Paper presented to the 25th
    Annual Convention, International Studies Association.
    Atlanta, Ga., March 22-26.

Cutler, Robert M. 1985a. "Domestic and Foreign Influences
    on Policy Making: The Soviet Union in the 1974 Cyprus
    Conflict." *Soviet Studies* **37**, no. 1 (January): 60-89.

Cutler, Robert M. 1985b. "Reporting Foreign News and
    Making Foreign Policy in the Soviet Union." Paper
    presented to the IIIrd World Congress for Soviet and
    East European Studies. Washington, D.C., October 30 -
    November 4.

Cutler, Robert M. 1987. "Harmonizing EEC-CMEA Relations: Never the Twain Shall Meet?" *International Affairs* (London) **63**, no. 2 (Spring): 259-270. Reprinted at pp. 365-381 in Diehl, 1989.

Cutler, Robert M. 1990. "Participation and Learning in Soviet Foreign Policy Making: The Press and Information-Processing under Brezhnev." Unpublished book manuscript.

Dallin, Alexander. "The Domestic Sources of Soviet Foreign Policy." Pp. 335-408 in Bialer, 1981.

Daniels, Robert Vincent. 1960. *The Conscience of the Revolution: Communist Opposition in Soviet Russia.* Cambridge: Harvard University Press.

Day, Richard B. 1981. *The "Crisis" and the "Crash": Soviet Studies of the West (1917-1939).* London: New Left Books.

Degras, Jane. 1967. "United Front Tactics in the Comintern, 1921-1928." Pp. 491-97 in Helmut Graber (comp.), *International Communism in the Era of Lenin.* Ithaca, N.Y.: Cornell University Press.

Deutsch, Karl W. 1963. *The Nerves of Government.* London: Free Press of Glencoe.

Diehl, Paul F. (ed.). 1989. *The Politics of International Organizations.* Chicago, Ill.: Dorsey.

Dzirkals, Lilita; Gustafson, Thane; and Johnson, A. Ross. 1982. *The Media and Intra-Elite Communication in the*

USSR. Rand Report R-2869. Santa Monica, Calif.: Rand Corporation, September.

Eran, Oded. 1973. "Soviet Foreign Policy—Random Institutional Observations." *International Problems* **23** (June): 82-94.

Eran, Oded. 1979. *The Mezhdunarodniki: An Assessment of Professional Expertise in the Making of Soviet Foreign Policy*. Jerusalem: Turtledove.

Etheredge, Lloyd S. 1981. "Government Learning: An Overview." Pp. 73-161 in *The Handbook of Political Behavior*, vol. 2, edited Samuel L. Long. New York: Plenum Press,

Fadeichev, E. M. 1971. "Istochniki informatsii: TASS i APN." In *Problemy informatsii v pechati: Ocherki teorii i praktiki*, edited by S. M. Gurevich, Moscow: Mysl'.

Fainsod, Merle. 1956. "Censorship in the USSR-A Documented Record." *Problems of Communism* **5**, no. 2 (March-April): 12-19.

Fainsod, Merle. 1963. "Bureaucracy and Modernization: The Russian and Soviet Case." Pp. 233-67 in *Bureaucracy and Political Development*, edited by Joseph LaPalombara. Princeton, N.J.: Princeton University Press.

Finn, A. 1954. *Experiences of a Soviet Journalist*. Mimeographed Series no. 66. New York: Research Program on the U.S.S.R.

Fleron, Frederic J., Jr. 1969a. "Co-optation as a Mechanism of Adaptation to Change: The Soviet Political Leadership System." *Polity* **2** (Winter): 176-201.

Fleron, Frederic J., Jr. (ed.). 1969b. *Communist Studies and the Social Sciences: Essays on Methodology and Empirical Theory.* Chicago, Ill.: Rand McNally.

Fleron, Frederic J., Jr. 1970. "Representation of Career Types in the Soviet Political Leadership." Pp. 108-39 in *Political Leadership in Eastern Europe and the Soviet Union*, edited by R. Barry Farrell. Chicago, Ill.: Aldine.

Fogelevich, L. G. (ed.). 1937. *Osnovye direktivy i zakonodatel'stvo o pechati*, 6th ed. Moscow: Ogiz.

Friedrich, Carl J., and Brzezinski, Zbigniew K. *Totalitarian Dictatorship and Autocracy,* (1st ed.). Cambridge: Harvard University Press.

Gaev, A. 1953a. "Kak delaetsia 'Pravda'". *Vestnik Instituta po izucheniiu istorii i kul'tury SSSR*, no. 4: 87-96

Gaev, A. 1953b. "'Pravda' and the Soviet Press." *Studies on the Soviet Union*, o.s., no. 1: 86-94.

Gaev, A. 1955. "Tsenzura sovetskoi pechati." *Issledovaniia i materialy* (Munich: Institute for the Study of the History and Culture of the U.S.S.R.), ser. 2, no. 2: 21-25.

Gantman, V. 1972. "Tekushchie problemy mirovoi politiki (15 sentiabria - 15 dekabria 1971 g.)." *Mirovaia ekonomika i mezhdunarodnye otnosheniia*, no. 1 (January): 72-90.

Garthoff, Raymond L. 1985. *Détente and Confrontation: American-Soviet Relations from Nixon to Reagan.* Washington, D.C.: Brookings Institution.

Glassman, Jon D. 1968. "Soviet Foreign Policy Decision-Making." Pp. 373-402 in *Columbia Essays in International Affairs: Volume III, The Dean's Papers, 1967.* New York: Columbia University Press.

Graboski, Tadeusz, and Nowak, Zdzislaw (eds.). 1969. *Integracja ekonomiczna Europy zachodniej i jej aspekty polityczno-militarne.* Poznan: Instytut zachodni.

Greenspan, Alan. 1997. "The Embrace of Free Markets." Speech to the Woodrow Wilson Award Dinner of the Woodrow Wilson International Center for Scholars. New York, June 10.

Griffiths, Franklyn. 1971. "A Tendency Analysis of Soviet Policy-Making." Pp. 335-78 in Skilling and Griffiths, 1971.

Griffiths, Franklyn. 1972. "Images, Politics, and Learning in Soviet Behaviour toward the United States." Ph.D. diss., Columbia University.

Griffiths, Franklyn. 1981. "Ideological Development and Foreign Policy." Pp. 19-48 in Bialer, 1981.

Griffiths, Franklyn. 1984. "The Sources of American Conduct: Soviet Perspectives and Their Policy Implications." *International Security* **9** (Fall): 3-50.

Griffiths, Franklyn. 1991. "Attempted Learning: Soviet Policy Toward the United States, in the Brezhnev Era." Pp. 630-83 in Breslauer and Tetlock, 1991.

Gromeka, V. 1974. "The United States-Western Europe: Scientific and Technological Competition." *International Affairs* (Moscow), no. 6 (June): 33-40.

Haas, Ernst B. 1991. "Collective Learning: Some Theoretical Speculations." Pp. 62-99 in Breslauer and Tetlock, 1991.

Heradstveit, Daniel, and Narvesen, Ove. 1978. "Psychological Constraints on Decision Making. A Discussion of Cognitive Approaches: Operational Code and Cognitive Map." *Cooperation and Conflict* **13**, no. 2: 77-92.

Hoffmann, Erik. P. 1968. "Communication Theory and the Study of Soviet Politics." *Canadian Slavic Studies* **2**, no. 4 (Winter): 542-58. Reprinted at pp. 379-98 in Fleron, 1969b.

Hoffmann, Erik P., and Fleron, Frederic J., Jr. (eds.). 1980. *The Conduct of Soviet Foreign Policy*, 2nd ed. New York: Aldine.

Hollander, Gayle Durham. 1972. *Soviet Political Indoctrination: Developments in Mass Media and Propaganda since Stalin*. New York: Praeger.

Hough, Jerry F. 1972. "The Soviet System: Petrification or Pluralism?" *Problems of Communism*, **21**, no. 2 (March-April): 25-45. Reprinted at pp. 19-48 in Hough, 1977.

Hough, Jerry F. 1977. *The Soviet Union and Social Science Theory*. Cambridge: Harvard University Press.

Hough, Jerry F. 1980. "The Evolution in the Soviet World View." *World Politics* **32**, no. 4 (July): 509-30.

Inozemtsev, N. N. 1972. "Les relations internationales en Europe dans les années 1980." Pp. 121-36 in *Europe 1980: L'avenir des relations intra-européennes; Rapports presentés à la Conférence des Directeurs et Représentants des Instituts européens de relations internationales, Varna, 3-5.X.1972.* Leiden: A. W. Sijthoff.

Karaganov, S. A. 1998. "Eshche neskol'ko soobrazhenii." *Mirovaia ekonomika i mezhdunarodnye otnosheniia*, no. 12 (December): 37-41.

Kennan, George F. 1967. *Memoirs, 1925-1950.* Boston, Mass.: Little, Brown.

Koschwitz, Hansjürgen. 1979. "Internationale Publizistik und Massenkommunication: Aufriss historischer Entwicklungslinien und gegenwärtiger Trends." *Publizistik* **24**, no. 4 (October-December): 458-83.

Kotlyar, A. 1955. *Newspapers in the USSR—Recollections and Observations of a Soviet Journalist.* Mimeographed Series no. 71. New York: Research Program on the U.S.S.R.

Kozlov, A. V. 1975. "Sovremennyi valiutnyi krizis i mezhimperialisticheskie protivorechiia (SShA-Zapadnaia Evropa)." Diss. kand. ekon. nauk, Moskovskii finansovyi institut.

Krasin, Iu. 1971. *Glavnaia revoliutsionnaia sila v tsitadeliakh kapitalizma.* Moscow: Znanie.

Krasin, Iu. 1974. "Sotsialisty Frantsii: Uroki istorii." Review of *Frantsuzskaia sotsialisticheskaia partiia v periode*

mezhdu dvumia mirovymi voinami, 1920-1940 gg., by
S. S. Salychev. *Mirovaia ekonomika i mezhdunarodnye
otnosheniia*, no. 4 (April): 143-46.

Kudriavtsev, V. 1974. "Tochka zreniia zhurnalista." *Zhurnalist*,
no. 12 (December): 29-31.

Kudriavtsev, Vik. 1974. "Vokrug 'kiprskogo krizisa'." *SShA*,
no. 9 (September): 72-75.

Kukhtevich, T. N. 1972. "O taktike edinstva deistvii
kommunistov s nemarksistskimi partiiami (na
opyte parlamentskoi bor'by proletariata razvitykh
kapitalisticheskikh stran Zapadnoi Evropy)." *Vestnik
Moskovskogo universiteta*, ser. 13, *Teoriia nauchnogo
kommunizma*, no. 2 (March-April): 51-60.

Kunze, Christine. 1978. *Journalismus in der UdSSR: eine
Untersuchung über Aufgaben und Funktionen sowjetischer
Journalisten unter besonderer Berucksichtigung der
Struktur der Massenmedien in der UdSSR und der
Diskussion des Berufsbildes in der Zeitung "Z[h]urnalist"*.
Munich: Verlag Dokumentation.

Legvold, Robert. 1974. "The Problem of European Security."
Problems of Communism **23** (January-February): 13-33.

Leites, Nathan. 1951. *The Operational Code of the Politburo*,
(1st ed.). New York: McGraw-Hill.

Lendvai, Paul. 1981. *The Bureaucracy of Truth: How
Communist Governments Manage the News*. Boulder,
Colo.: Westview.

Lewin, Kurt. 1949. "Cassirer's Philosophy of Science and the Social Sciences." In *The Philosophy of Ernst Cassirer*, edited by Paul Arthur Schipp. Library of Living Philosophers 49. Evanston Ill.: Northwestern University Press.

Lodge, Milton. 1969. *Soviet Elite Attitudes since Stalin*. Columbus, Ohio: C.E. Merrill.

Löwenhardt, John. 1981. *Decision Making in Soviet Politics*. New York: St. Martin's Press.

Löwenthal, Richard. 1970. "Development vs. Utopia in Communist Policy." Pp. 33-116 in *Change in Communist Systems*, edited by Chalmers Johnson. Stanford, Calif.: Stanford University Press.

Marcuse, Herbert. 1960. *Reason and Revolution: Hegel and the Rise of Social Theory*. Boston, Mass.: Beacon Press.

Matlock, Jack M., Jr. 1995. *Autopsy on an Empire: The American Ambassador's Account of the Collapse of the Soviet Union*. New York: Random House.

Meissner, Boris. 1977. "Der Auswärtige Dienst der UdSSR." *Aussenpolitik* **28**, no. 1: 47-61.

Mel'nikov, D. 1972. "Zapadnoevropeiskii tsentr imperializma." *Mirovaia ekonomika i mezhdunarodnye otnosheniia*, no. 1 (January): 14-30.

Meyer, Alfred G. 1965. *The Soviet Political System*. New York: Knopf.

Naylor, Thomas H. 1988. *The Gorbachev Strategy: Opening the Closed Society*. Lexington, Mass.: D. C. Heath, Lexington Books.

Peffley, Mark A., and Hurwitz, Jon. 1985. "A Hierarchical Model of Attitude Constraint." *American Journal of Political Science* **29**, no. 4 (November): 871-90.

Pel't, V. D. 1980. "Sobstvennyi i spetsial'nyi korrespondenty." Pp. 74-87 in *Teoriia i praktika sovetskoi periodicheskoi pechati*, edited by Pel't. Moscow: Vysshaia shkola.

Pethybridge, Roger. 1962. *A Key to Soviet Politics: The Crisis of the Anti-Party Group*. New York: Praeger.

Petrov, Vladimir. 1973. "Formation of Soviet Foreign Policy." *Orbis* **17**, no. 3 (Fall): 819-50.

Popov, V. P. 1981. "Politicheskie obozrevateli (mezhdunarodniki) v massovo-politicheskoi gazete." Diss. kand. filol. nauk, Moskovksii gosudarstvennyi universitet.

Popov, V. P. 1984. *Na glavnom napravlenii (O rabote politicheskogo obozrevatelia)*. Moscow: Mysl'.

Primakov, E. 1987. "Novaia filosofiia vneshnei politiki." *Pravda*, July 10, p. 4.

Rasputins, B. I. 1980. *Sovetskaia istoriografiia sovremennogo rabochego dvizheniia*. Chast' 2, *Sotsial'no-politicheksie problemy*. L'vov: Vishcha shkola, Izdatel'stvo pri L'vovskom gosudarstvennom universitete.

Remington Thomas F. 1985a. "Politics and Professionalism in Soviet Journalism." *Slavic Review* **44**, no. 3 (Fall): 489-503.

Remington, Thomas F. 1985b. *The Truth of Authority: Ideology and Communication in the Soviet Union.* Pittsburgh, Penna.: University of Pittsburgh Press.

Rémy, Alain. 1981. "Le rôle de l'or dans l'économie monétaire occidentale: Analyses soviétiques." Thèse pour le doctorat de troisième cycle, Université de Paris I—Panthéon-Sorbonne, October.

Révész, László. 1974. *Recht und Willkür in der Sowjetpresse: eine presserechtliche und pressepolitische Untersuchung.* Freiburg: Universitätsverlag.

Révész, László. 1975. *Moskau am Atlantik? Indizien für eine kommunistische Machtergreifung in Portugal.* SOI-Sonderdruck 6. Bern: Schweizerischer Ost-Institut.

Révész, László. 1976. *Moskau über Portugal: Taktische Fragen und Medienpolitik.* SOI-Sonderdruck 13. Bern: Schweizerischer Ost-Institut.

Romanchuk, E. F. 1970. "Redaktsionnye kadry respublikanskikh gazet." Pp. 328-45 in *Voprosy teorii i praktiki massovykh sredstv propagandy.* Vyp. 3, edited by V. D. Datsiuk. Moscow: Mysl', 1970.

Rosenfeldt, Niels Erik. 1978. *Knowledge and Power: The Role of Stalin's Secret Chancellery in the Soviet System of Government.* Copenhagen: Rosenkilde and Bagger.

Roth Paul. 1980. *Sow-Inform: Nachrichtenwesen und Informationspolitik der Sowjetunion*. Düsseldorf: Droste Verlag GmbH.

Roth, Paul. 1982. *Die kommandierte öffentliche Meinung: Sowjetische Medienpolitik*. Stuttgart: Seewald.

Rühl, Manfred. 1969. *Die Zeitungsredkation als organisiertes sociales System*. Bielefield: Bartelsmann Universitätsverlag.

Salychev, S. S. 1971. *Problemy revoliutsionnoi bor'by v stranakh kapitala na sovremennom etape*. Moscow: Znanie.

Schapiro, Leonard. 1975. "The General Department of the CC of the CPSU," *Survey*, no. 96: 53-65.

Schapiro, Leonard. 1977. "The International Department of the CPSU: Key to Soviet Policy." *International Journal* **32**, no. 1 (Winter): 41-55.

Sekerin, V. P. 1973. "Kontent-analiz v kompleksnom izuchenii gazety (Nekotorye metodicheskie vyvody)." Pp. 58-62 in *Metodologicheskie i metodicheskie problemy kontent-analiza (Tezisy dokladov rabochego soveshchaniia sotsiologov)*. Vyp. 2, edited by A. G. Zdravomyslov. Moscow-Leningrad: Institut sotsiologicheskikh issledovanii AN SSSR, 1973.

Semeiko, L. S. 1988. "SS-20: Oshibka, no men'shaia, chem mozhno bylo by dumat'." *Mirovaia ekonomika i mezhdunarodnye otnosheniia*, no. 12 (December): 32-36.

Shevardnadze, E. A. 1987. "V Ministerstve inostrannykh del SSSR: Vystuplenie E. A. Shevardnadze na sobranii aktiva Diplomaticheskoi akademii, Instituta mezhdunarodnykh otnoshenii i tsentral'nogo apparta MID SSSR, 27 iunia 1987." *Vestnik Ministerstva inostrannykh del SSSR*, no. 2: 30-34.

Shkondin, M. V. 1982. *Pechat': Osnovy organizatsii i upravleniia*. Moscow: Izdatel'stvo MGU.

Shulman, Marshall D. 1948. "The Administration of the Soviet Press." M.A. thesis, Columbia University, The Russian Institute. Cited by permission of the author.

Skilling, H. Gordon. 1966. "Interest Groups and Communist Politics." *World Politics* **18**, no. 3 (April): 435-51.

Skilling, H. Gordon. 1983. "Interest Groups and Communist Politics Revisited." *World Politics* **36**, no. 1 (October): 1-27.

Skilling, H. Gordon, and Griffiths, Franklyn (eds.). 1971. *Interest Groups in Soviet Politics*. Princeton, N.J.: Princeton University Press.

Snyder, Richard C.; Bruck, H. W.; and Sapin, Burton. 1962. "Decision-Making as an Approach to the Study of International Politics." Pp. 14-185 in *Foreign Policy Decision Making: An Approach to the Study of International Politics*, edited by Snyder, Bruck, and Sapin. New York: Free Press of Glencoe.

Solomon, Peter H., Jr. 1978. *Soviet Criminologists and Criminal Policy: Specialists in Policy-Making*. New York: Columbia University Press.

Stadnichenko, A. 1971. "Demonetizatsiia zolota ne proiskhodit." *Mirovaia ekonomika i mezhdunarodnye otnosheniia*, no. 10 (October): 93-102.

Stankiewicz, W. 1972. "The Contradictions of Military-Industrial Integration in Western Europe." *International Affairs* (Moscow), no. 7 (July): 28-34.

Steinbruner, John D. 1974. *The Cybernetic Theory of Decision: New Dimensions of Political Analysis.* Princeton, N.J.: Princeton University Press.

Stepanov, L. 1972. "Tekushchie problemy mirovoi politiki (15 dekabria 1971 g. - 15 marta 1972 g.)." *Mirovaia ekonomika i mezhdunarodnye otnosheniia*, no. 4 (April): 69-88.

Sturua, G. M. 1988. "Bylo li neobkhodimo razvertyvanie raket SS-20?" *Mirovaia ekonomika i mezhdunarodnye otnosheniia*, no. 12 (December): 23-28.

Troianskii, M. G. 1980. "Propagandistskaia deiatel'nost' portugal'skoi kommunisticheskoi partii v zashchitu i uglublenie osnovnykh zavoevanii aprel'skoi revoliutsii 1974 goda (aprel' 1974 - noiabr' 1976 gg.)." Diss. kand. ist. nauk, Akademiia obshchestvennykh nauk pri TsK KPSS.

Tsukasov, S. V. 1973. "Osnovnye tendentsii razvitiia organizatsii raboty redaktsii tsentral'noi gazety." Pp. 5-18 in *Problemy nauchnoi organizatsii zhurnalistskogo truda i raboty redaktsionnykh kollektivov (Materialy I Vsesoiuznoi nauchno-prakticheskoi konferentsii, 25-26 maia 1972 g.).* Moscow: Izdatel'stvo MGU.

Tsukasov, S. V. 1975. *Vremia zrelosti.* Moscow: Mysl'.

Valkenier, Elizabeth Kridl. 1983. *The Soviet Union and the Third World: An Economic Bind.* New York: Praeger.

Vil'chek, Vsevolod. 1973. "Professionaly i diletanty." *Zhurnalist*, no. 9 (September): 16-18.

Wettig, Gerhard. 1975. *Die sowjetische Portugal-Politik, 1974-1975.* Bericht 60/1975. Cologne: Bundesinstitut für ostwissenschaftliche und internationale Studien.

Winch, Peter. 1958. *The Idea of a Social Science.* London: Routledge and Kegan Paul.

Wittkämpfer, Gerhard W., and Bellers, Jürgen. 1986. "The Press and Foreign-Policy Decision Making: An Analysis of German-Polish Negotiations in 1969-1970." *International Political Science Review* **7**, no. 4: 400-414.

Zagoria, Donald S. 1962. *The Sino-Soviet Conflict 1956-1961.* Princeton, N.J.: Princeton University Press.

Zak, L. A. 1976. *Zapadnaia diplomatiia i vneshnepoliticheskie stereotipy.* Moscow: Mezhdunarodnye otnosheniia.

Zaretskii, V. A. 1973. "Vzaimosviaz' demokraticheskikh i sotsialisticheskikh zadach na sovremennom etape klassovoi bor'by v razvitykh kapitalisticheskikh stranakh". Pp. 102-19 in *Problemy rabochego, kommunisticheskogo i natsional'no-osvoboditel'nogo dvizheniia (Nauchnye trudy),* edited by A. A. Koval'skii and M. M. Solntseva. Moscow: MGIMO, 1973.

Zhdanova, L. P. 1971. "Valiutnaia sistema kapitalizma." Pp. 63-68 in *Ekonomicheskoe polozhenie kapitalisticheskikh i razvivaiushchikhsia stran: Obzor za 1970 g. i nachalo 1971 g.* Moscow: Pravda.

Zhukov, Iurii. 1980 "K chitateliu." Pp. 3-12 in Georgii Ratiani, *Na Blizhnem i Dal'nem Zapade: Kniga vtoraia.* Moscow: Pravda.

Zimmerman, William. 1969. *Soviet Perceptions of International Relations, 1956-1967.* Princeton, N.J.: Princeton University Press.

Znamenskaia, T. Iu., and Smagin, A. V. 1988. "Vneshnepoliticheskaia propaganda: Voina slov ili zainteresovannyi dialog?" *SShA*, no. 9 (September): 22-33.

About the author

Dr. Robert M. Cutler (http://www.robertcutler.org), is senior research fellow in the Institute of European, Russian and Eurasian Studies, Carleton University, Canada. Educated at the Massachusetts Institute of Technology, Graduate Institute of International Studies (Geneva) and The University of Michigan, he has held teaching positions and research fellowships at major universities in the United States, Canada, France, Switzerland and Russia. His scholarly articles have appeared in the leading academic and policy journals of several fields.

He has been writing on Soviet and Eurasian affairs for a third of a century. Cutler has won numerous competitive grants and served on a half-dozen academic-journal and policy-review editorial boards in addition to executive committees of professional scholarly and policy research organizations. He maintains a strong Internet and media presence through analytical commentaries on contemporary affairs, notably international relations, energy security and geo-economics, including Asian finance. In addition, he consults with international institutions, NGOs, think tanks, governments and the private sector both in the energy field and on questions of organizational design and decision-making analysis in the complexity-science perspective.

www.ingramcontent.com/pod-product-compliance
Lightning Source LLC
Chambersburg PA
CBHW070811280326
41934CB00012B/3149